T0214916

SpringerBriefs on PDEs and Data Science

SpringerBriefs on PDEs and Data Science targets contributions that will impact the understanding of partial differential equations (PDEs), and the emerging research of the mathematical treatment of data science.

The series will accept high-quality original research and survey manuscripts covering a broad range of topics including analytical methods for PDEs, numerical and algorithmic developments, control, optimization, calculus of variations, optimal design, data driven modelling, and machine learning. Submissions addressing relevant contemporary applications such as industrial processes, signal and image processing, mathematical biology, materials science, and computer vision will also be considered.

The series is the continuation of a former editorial cooperation with BCAM, which resulted in the publication of 28 titles as listed here: https://www.springer.com/gp/mathematics/bcam-springerbriefs

Avy Soffer • Chris Stucchio • Minh-Binh Tran

Time Dependent Phase Space Filters

A Stable Absorbing Boundary Condition

 Springer

Avy Soffer
Department of Mathematics
Rutgers, The State University of New Jersey
Piscataway, NJ, USA

Chris Stucchio
Department of Mathematics
Rutgers, The State University of New Jersey
Piscataway, NJ, USA

Minh-Binh Tran
Department of Mathematics
Texas A&M University
College Station, TX, USA

ISSN 2731-7595   ISSN 2731-7609   (electronic)
SpringerBriefs on PDEs and Data Science
ISBN 978-981-19-6817-4   ISBN 978-981-19-6818-1   (eBook)
https://doi.org/10.1007/978-981-19-6818-1

Mathematics Subject Classification: 35Q41, 35Q55, 35S05, 65N30, 65N35

This Springer imprint is published by the registered company Springer Nature Singapore Pte Ltd.
The registered company address is: 152 Beach Road, #21-01/04 Gateway East, Singapore 189721,
Singapore

Preface

Many fundamental laws of nature are formulated by Partial Differential Equations (PDEs) that model the propagation of waves in space. Among them are quantum waves, light and acoustic waves, gravitational waves and many others. The applications of PDEs to science, engineering and real-life situations, as well as theoretical physics investigations necessitate numerical approximation schemes.

As waves propagate, we need to solve the PDEs on large domain for a large time; it is then not practical to have the domain of computation growing with time. Therefore, in actual computations, it is necessary to introduce artificial boundary terms that absorb the spreading waves. Such absorbing boundary conditions were not perfect, as they were typically based on trial and error, by adding sinks to the domain of computation.

A revolutionary new idea was introduced in 1977 by Enquist and Majda. Being inspired by Scattering Theory, they proposed a completely new way of constructing the absorbing boundaries, by projecting out *only the outgoing waves reaching the boundary*. The idea was implemented in a rigorous way for the case of the wave equation and has become the leading tool in the numerics of hyperbolic systems of equations. It led to the development of more refined constructions, which could also be used for other dispersive equations, and is known as Perfectly Matched Layer (PML). As one of the authors (A.S.) learned from Majda, this approach is not working well for general dispersive equations, including those that originate from Quantum Mechanics. The reasons for that are well known: The uncertainty principle makes it impossible to have a local (in space) decomposition of waves into incoming and outgoing waves. In the case of the wave equation, it is very simple and local: the operators $\partial_x \pm \partial_t$ annihilate the incoming/outgoing part of the solution of the wave equation (in flat space). While there are various methods to project on incoming/outgoing waves for dispersive equations, which were used effectively in scattering theory, these constructions are highly non-local. The lack of accuracy in these types of filtering causes instabilities in numerical computations, in particular in cases when the group velocity of the wave is pointing in the opposite direction to the wave number.

This book is devoted to a new approach to filtering of the outgoing waves at the boundary layer. The approach is of general nature, and works for dispersive and hyperbolic equations. We will mostly describe the method and the ideas behind it for the case of the Schrödinger equation. The approach is stable by construction, versatile and rigorously controlled. The decomposition of the solution into incoming and outgoing waves at a boundary layer is performed by a new approach, in which the solution in the boundary is expanded in terms of a *frame* in which the elements are Gaussians of a fixed width, times a phase factor of the form $e^{ix \cdot v}$. Each Gaussian corresponds to a coherent state of the Schrödinger equations, which is centered at some point x_0 and is moving (under the free flow) with a velocity v. The Gaussian is outgoing (and therefore removed) if the motion in the direction v starting at the point x_0 moves out of the domain of computation. Using such filtering periodically in time, one guarantees the stabilityu of the outgoing waves. The errors in this approach are computable, as they correspond to Gaussians which have tangential velocities to the boundary. By knowing the lowest and highest relevant group velocities, it is possible to optimize the lattice spacing in the x_0, v variables, for a given accuracy.

This approach is then augmented with multi-scale techniques to deal with the formation of long wavelength slow oscillations that may appear over time. Our approach can be employed to various shapes of domains, while the scheme used to solve the equation is not restricted to any particular method. The construction of the filtering, the rigorous analysis to control the errors, proof of optimality of the approach and numerical examples are presented.

Piscataway, NJ, USA   Avy Soffer
Piscataway, NJ, USA   Chris Stucchio
College Station, TX, USA   Minh-Binh Tran

Acknowledgment

Avy Soffer is supported in part by Simons Foundation Grant number 851844 and NSF-DMS-2205931. Minh-Binh Tran is funded in part by a Humboldt Fellowship, NSF CAREER DMS-2044626/DMS-2303146, NSF Grants DMS-1854453, DMS-2204795, DMS-230552.

Contents

Chapter 1
Introduction

1.1 The Open Boundary Problem

Consider a semilinear Schrödinger equation on $\mathbb{R}^{N+1}$

$$i\partial_t \Psi(\vec{x}, t) = -(1/2)\Delta\Psi(\vec{x}, t) + g(t, \vec{x}, \Psi(\vec{x}, t))\Psi(\vec{x}, t) \qquad (1.1.1)$$

where $g(t, \vec{x}, \cdot)$ is some semilinear, Lipschitz (in some Sobolev space) nonlinearity. For instance, $g(t, \vec{x}, \cdot)$ could be $V(\vec{x}, t) + f(|\Psi(\vec{x}, t)|^2)$ for some smooth function f and (spatially) localized potential $V(\vec{x}, t)$.

We assume the initial condition and nonlinearity are such that the nonlinearity remains localized inside some box $[-L_{NL}, L_{NL}]^N$. Outside this region the solution is assumed to behave like a free wave (a solution to (1.1.1) with $g(t, \vec{x}, \cdot) = 0$), which is well understood.

One very common method of solving such a problem is domain truncation. That is, one solves the PDE (1.1.1) numerically on a region $[-L, L]^N$. On the finite domain, of course, boundary conditions must be specified. Dirichlet and Neumann boundaries introduce spurious reflections, while periodic boundaries (which are naturally imposed by spectral methods) allow outgoing waves to wrap around the computational domain. In either case, a serious mistake has been made. This causes the numerical solution to become incorrect after a time $T \approx L/k_{sup}$, where k_{sup} is the "maximal velocity" of the solution.[1]

It is an interesting and well known problem to find a way to minimize these errors. The simplest way is simply to expand the domain as the support of $\Psi(\vec{x}, t)$ grows, but this is computationally very expensive.

---

[1] There is, in general, no maximal velocity of the solution. However, we will define k_{sup} more precisely later on. k_{sup} will be the frequency such that the energy of $\hat{\Psi}(k, t)$ for $k > k_{sup}$ is small.

© The Author(s), under exclusive license to Springer Nature Singapore Pte Ltd. 2023
A. Soffer et al., *Time Dependent Phase Space Filters*, SpringerBriefs on PDEs
and Data Science, https://doi.org/10.1007/978-981-19-6818-1_1

For the wave equation and other strictly hyperbolic wave equations this problem has a beautiful exact solution (c.f. [32, 33]), namely the Dirichlet-to-Neumann map. The Eq. (1.1.1) is solved in a region $[-L_{\text{in}}, L_{\text{in}}]^N$, and the boundary conditions are given by $\Psi(\vec{x}, t)$ (where $\Psi(\vec{x}, t)$ is the solution to (1.1.1) on $\mathbb{R}^N$) on the boundary. Of course, since $\Psi(\vec{x}, t)$ is not known, it must be approximated. The usual method (used with great success for the wave equation) is to approximate the exact solution by rational functions in the frequency domain. These correspond to boundary conditions given by a high order differential operator. This result depends strongly on the fact that in the frequency domain, the Dirichlet-to-Neumann map behaves like a polynomial at ∞.

For the Schrödinger equation and other dispersive wave equations, the situation is not so simple. Even in the free case ($g(t, \vec{x}, \Psi(\vec{x}, t)) = 0$), it may be impractical to construct local (in time and space) approximations to the Dirichlet-Neumann operator for large time interval as the expansions required may need a huge number of terms. In addition, constructing the Dirichlet-Neumann map in the case $-(1/2)\Delta + V(x)$ is not an easy matter. In the nonlinear case we know of results only in 1 space dimension, and with no rigorous error estimates [66, 69, 70].

Another drawback of the Dirichlet-to-Neumann approach is that it precludes the use of Fourier transform methods to solve the interior problem. Fourier transform methods (described on p. 52) use the FFT (Fast Fourier Transform) to diagonalize the operator $e^{i(1/2)\Delta t}$. This approach naturally imposes periodic boundaries. Fourier transform methods are desirable, since they are believed to be more accurate than most other methods on periodic domains (for a finite set of spatial frequencies). The error due to boundary conditions, however, makes them unfeasible. Thus, one usually reverts to using FDTD (Finite Difference Time Domain) in their place, but the accuracy of these methods is limited and decreases rapidly with high spatial frequencies.

An ad-hoc approach (described in, e.g. [51]), the current industry standard, is to add an absorbing potential, $-iV(x)\Psi(\vec{x}, t)$ to the right hand side of (1.1.1), with $V(x) = 0$ away from the boundary. This potential has the effect of partially dissipating waves as they pass over it. Thus, as waves reach the boundary, they are partially dissipated by the complex potential, reducing the reflection. This approach is far from optimal, but is still the industry standard due to the ease of implementation, compatibility with spectral methods and simplicity.

A variant on this approach is the PML (Perfectly Matched Layer). Proposed originally for Maxwell's equations in [18] and for the Schrödinger equation in [37], it is a variant on the absorbing potential method in which Δ is replaced by $(1 - ia(x))\Delta$ (with $a(x)$ nonzero only in a boundary layer) in such a way so that when $a(x)$ "switches on", there is no reflection at the interface.

1.1.1 Filtering Approach

We present an alternative approach to absorbing boundaries, first described in [62]. Our method differs in a fundamental way from previous approaches: it is, as the name points out, a filter on the boundary of the domain. It removes the outgoing waves that reach the boundary. As such, we have a complete explicit knowledge of the removed part (and can propagate it outside the domain by the free flow, for example.) Furthermore, it is possible to introduce incoming waves as a source into the domain through the boundary. This allows the analysis of hybrid quantum-classical domain calculations (nano-device coupled to a classical current source). The way the filtering is done is insensitive to the shape of the boundary. We make the assumption that near the boundary of the box, the solution behaves like a free wave. We make no assumptions on the nonlinearity, beyond the fact that it is localized on the inside of the box and locally Lipschitz. In particular, the nonlinearity could take the form of a complicated time dependent short range potential $V(\vec{x}, t)\Psi(\vec{x}, t)$, a polynomial nonlinearity $f(|\Psi(\vec{x}, t)|^{2\sigma})\Psi(\vec{x}, t)$ (for $f(z)$ a Lipschitz function) or others.

We also assume that the solution remains localized in frequency, that is $\hat{\Psi}(\vec{k}, t)$ is small for $k \notin [-k_{\text{sup}}, k_{\text{sup}}]^N$ for some large number k_{sup} (the maximal momentum), a common assumption to make for numerical purposes.

We further assume the existence of a low frequency k_{inf} with the property that $\hat{\Psi}(\vec{k}, t)$ has little mass on the region $\{\vec{k} : |\vec{k}| < k_{\text{inf}}\}$. This assumption is very difficult to verify, and is not always true; an algorithm capable of treating the case where the mass near $k = 0$ is unbounded is detailed in [63]. For the algorithm to be efficient, it should be the case that $k_{\text{inf}}^{-1} \ll L_{\text{in}}$, i.e. the interaction region is much longer than the smallest wavelength.

Our algorithm is as follows. We assume the initial data is localized on a region $[-L_{\text{in}}, L_{\text{in}}]^N$. We solve (1.1.1) on the box $[-(L_{\text{in}} + w), L_{\text{in}} + w]^N$ on the time interval $[0, T_{\text{st}}]$.

By making T_{st} small enough (smaller than w/k_{sup}), we can ensure that $\Psi(\vec{x}, t)$ is mostly localized inside box $[-(L_{\text{in}} + w), (L_{\text{in}} + w)]^N$. Thus, since very little mass has actually hit the boundaries, very little has reflected and we have made few errors.

We now decompose the solution $\Psi(\vec{x}, t)$ into a sum of gaussians (indexed by $\vec{a}, \vec{b} \in \mathbb{Z}^N$, with x_0, k_0, σ all positive constants satisfying certain constraints to be made precise later):

$$\Psi(x, T_{\text{st}}) = \sum_{(\vec{a}, \vec{b}) \in \mathbb{Z}^N \times \mathbb{Z}^N} \Psi_{(\vec{a}, \vec{b})} \pi^{-N/4} \sigma^{-N/2} e^{ik_0 \vec{b} \cdot \vec{x}} e^{-|\vec{x} - \vec{a}x_0|_2^2 / 2\sigma^2}$$

Because $\hat{\Psi}(\vec{k}, t)$ is localized on $[-k_{\text{sup}}, k_{\text{sup}}]^N$, $\Psi(nT_{\text{st}})_{(\vec{a}, \vec{b})} \approx 0$ is for $|\vec{b}k_0|_\infty > k_{\text{sup}}$. Because $\Psi(\vec{x}, t)$ is localized on $[-(L_{\text{in}} + w), (L_{\text{in}} + w)]^N$, $\Psi_{(\vec{a}, \vec{b})} \approx 0$ for $|\vec{a}x_0|_\infty > L_{\text{in}} + w$. Therefore:

$$\Psi(x, T_{\text{st}}) \approx \sum_{\substack{|\vec{a}x_0|_\infty \leq L_{\text{in}} + w \\ |\vec{b}k_0|_\infty \leq k_{\text{sup}}}} \Psi_{(\vec{a}, \vec{b})} \pi^{-N/4} \sigma^{-N/2} e^{ik_0 \vec{b} \cdot \vec{x}} e^{-|\vec{x} - \vec{a}x_0|_2^2 / 2\sigma^2}$$

We then examine the gaussians near the boundary (with $|\vec{a}x_0|_\infty \geq L$) and determine whether they are leaving the box or not (after propagation under the free flow). This is simple enough to do, since elementary quantum mechanics tells us that:

$$e^{i(1/2)\Delta t} \pi^{-N/4} \sigma^{-N/2} e^{ik_0 \vec{b} \cdot \vec{x}} e^{-|\vec{x} - \vec{a}x_0|_2^2 / 2\sigma^2}$$

$$= \frac{\exp(i\vec{b}k_0 \cdot (\vec{x} - \vec{b}k_0/2t - \vec{a}x_0))}{\pi^{N/4} \sigma^{N/2} (1 + it/\sigma^2)^{N/2}} \exp\left(\frac{-|\vec{x} - \vec{b}k_0 t - \vec{a}x_0|_2^2}{2\sigma^2 (1 + it/\sigma^2)}\right)$$

Essentially, $e^{i(1/2)\Delta t} \pi^{-N/4} \sigma^{-N/2} e^{ik_0 \vec{b} \cdot \vec{x}} e^{-|\vec{x} - \vec{a}x_0|_2^2 / 2\sigma^2}$ moves along the trajectory $\vec{a}x_0 + \vec{b}k_0 t$, while spreading about it's center at the rate σ^{-1}.

Then, if a given gaussian is leaving the box, we delete it. If it is not, we keep it. Gaussians with low velocities ($|\vec{b}k_0| \leq k_{\text{inf}}$) spread laterally more quickly than their center of mass moves, and are thus both incoming and outgoing. In this work we assume that these gaussians have small mass, and merely quantify the errors caused by their presence. This is done by assuming the existence of k_{inf}, and choosing $w \geq O(k_{\text{inf}}^{-1})$.

After this filtering operation, the only gaussians remaining are either inside the box $[-L_{\text{in}}, L_{\text{in}}]^N$, or inside the box $[-(L_{\text{in}} + w), (L_{\text{in}} + w)]^N$ but moving towards $[-L_{\text{in}}, L_{\text{in}}]^N$. We then repeat the process, and propagate with periodic boundaries until $2T_{\text{st}}$, and filter again at this time.

This description is vague, and the rest of this paper fills in the details and provides theoretical justification of the method. In particular, we prove rigorous error bounds, subject to some relatively general assumptions (most of which can be estimated apriori or verified aposteriori). That is, for $t \in [0, T_{\mathfrak{M}}]$ (where $T_{\mathfrak{M}}$ is some maximal time interval of interest) we show that:

$$\sup_{t \in [0, T_{\mathfrak{M}}]} \left\| \chi_{[-L_{\text{in}}, L_{\text{in}}]^N}(x) \left(\Psi(\vec{x}, t) - \Phi(\vec{x}, t)\right) \right\|_{H^s} \leq \tau$$

where $\Phi(\vec{x}, t)$ is our approximate solution, τ is some prescribed error, and $H_b^s = H^s([-L_{\text{in}}, L_{\text{in}}]^N)$ a Sobolev space, with $s = 0, 1$. We believe that similar results can be proved for $s > 1$, although some calculations will be different (see Remark 6.1.4 and the exact calculations in Sect. 6.2, see also Remark 6.2.2 for an explanation of the modifications necessary for $s > 1$).

1.1.2   Method Accuracy

We calculate the error made at each step in the above analysis and then add it all up to get the global error bound.

For a general time-stepping algorithm (with periodic boundaries and no filtering), the error bound would take the following form:

$$\sup_{t \in [0, T_{\mathfrak{M}}]} \| \mathfrak{U}(t) \Psi_0(x) - \Phi(x, t) \|_{H_b^s} \leq \text{BoundaryError}(T_{\mathfrak{M}})$$

$$+ \text{HighFrequency}(T_{\mathfrak{M}}) + \text{LowFrequency}(T_{\mathfrak{M}})$$

$$+ \text{NonlocalNonlinearity}(T_{\mathfrak{M}}) + \text{Instability}(T_{\mathfrak{M}}) \qquad (1.1.2)$$

The term BoundaryError($T_{\mathfrak{M}}$) encompasses errors due to waves wrapping/reflecting from the boundaries of the box. For many problems, this is the dominant error term. It is directly proportional to the mass which would have (if we were solving the problem on $\mathbb{R}^N$) radiated outside the box $[-L_{\text{in}}, L_{\text{in}}]^N$.

The HighFrequency($T_{\mathfrak{M}}$) part stems from waves with momenta too high to be resolved by the discretization. The term LowFrequency($T_{\mathfrak{M}}$) encompasses errors due to waves with wavelength that is long in comparison to the box. The term NonlocalNonlinearity($T_{\mathfrak{M}}$) stems from that fraction of the nonlinearity itself which is located outside the box. The Instability($T_{\mathfrak{M}}$) stems from the possibility that the dynamics of the solution itself might amplify the other errors dramatically (e.g. in strongly nonlinear problems).

Our algorithm reduces the term BoundaryError($T_{\mathfrak{M}}$) only. We show, by a discrete variant of the gaussian beam method, that if we filter off the outgoing waves in the manner described previously, the boundary error term can be made arbitrarily small. The cost is increasing the width of the region in which filtering takes place.

The main drawback of our algorithm is that it does not provide us the ability to filter low frequency outgoing waves, that is to say waves for which the wavelength is longer than the buffer region. This is precisely what we would expect from the Heisenberg uncertainty principle.

Since the goal of this work is to reduce the error due to boundary reflection, all the error terms besides the boundary error term are made small by assumption. We provide no bounds on them, since these bound would depend very strongly on the specific form of $g(t, \vec{x}, \Psi(\vec{x}, t))\Psi(\vec{x}, t)$.

Remark 1.1.1 At first glance, it would appear that an absorbing boundary layer (either complex potential or PML) would reduce the boundary error nearly to zero, with the error being nothing more than those waves which it fails to absorb. This intuition is false, and a counterexample is provided in Sect. 11.2.

The reason is as follows. Suppose we add an absorbing boundary layer (denoted by A) term to (1.1.1). Let $\Psi_a(x, t)$ solve:

$$i \partial_t \Psi_a(x, t) = (-(1/2)\Delta + A)\Psi_a(x, t) + g(t, \vec{x}, \Psi_{a,b}(\vec{x}, t))\Psi_{a,b}(\vec{x}, t)$$

Let $\Psi_{a,b}(x,t)$ solve the corresponding periodic problem:

$$i\partial_t \Psi_{a,b}(x,t) = (-(1/2)\Delta_b + A)\Psi_{a,b}(x,t) + g(t,\vec{x},\Psi_{a,b}(\vec{x},t))\Psi_{a,b}(\vec{x},t)$$

It is true that $\left\| \Psi_a(x,t) - \Psi_{a,b}(x,t) \right\|_{H_b^s}$ is small (that is, the box problem with an absorber approximates the $\mathbb{R}^N$ problem with an absorber). However, it is not necessarily true that $\| \Psi(x,t) - \Psi_a(x,t) \|_{H_b^s}$ is small, because the $\mathbb{R}^N$ problem with an absorber may not accurately approximate the $\mathbb{R}^N$ problem with no absorber.

The TDPSF algorithm sidesteps this issue by directly approximating the solution on $\mathbb{R}^N$, and only using the box propagator on regions of phase space where it is guaranteed to be accurate.

1.1.3 The Advantages of the Method

Our method is versatile and general, in the sense that it is a numerical application of the gaussian beam method. Extensions and modifications to other sorts of equations are likely to be straightforward, although one might prefer to decompose $\Psi(\vec{x},t)$ into some other functions different than gaussians.[2]

In particular, we believe this can be extended without much difficulty to the free wave equation, replacing gaussians by curvelets [22, 23].

In addition, if the dynamics on the boundary are non-free, we believe our method can be modified to treat these dynamics effictively. Suppose that instead of propagating along the trajectory $\vec{a}x_0 + \vec{b}k_0 t$, a typical gaussian propagated along the trajectory $\gamma(\vec{a},\vec{b},t)$ instead. We could still apply our method, except now we would attempt to determine whether $\gamma(\vec{a},vb,t)$ is leaving the box rather than $\vec{a}x_0 + \vec{b}k_0 t$ when determining which gaussians to filter. We have no rigorous error bounds on this method at this time, however we believe they could be constructed by methods similar to what we do in this work.

Another advantage to our method is that when it does fail, it fails gracefully. The main mode of failure is for too many gaussian's to fall into the region where we cannot determine whether they are incoming or outgoing.[3] However, if this occurs, the algorithm is aware of it and an exception is raised. In addition, low frequency gaussians can be treated by an extension of this method as described in [63].

We expect proofs of the error bound in cases like this to be simple (albeit long) variations on the proof we give here.

---

[2] More precisely, for a given equation, one should use a family of coherent states which is also a frame. In addition, the family of coherent states should not make computations too complex.

[3] The other mode of failure is spectral blocking in the frequency domain, a common mode of failure for spectral methods. This problem occurs when the lattice spacing δx is too large to resolve the high frequencies generated by the problem.

1.1.4 The Disadvantages of the Method

Our method is based strongly on two main assumptions, which will not hold for every equation or every initial condition.

The most important assumption is that the solution behaves like a free wave outside of a certain box $[-L_F, L_F]^N$, and we demand that the computational region encompass this box. If this does not hold, the error bound we provide is no longer valid. An example of this is the case of a moving soliton which leaves the box.[4] The dynamics near the boundary are no longer free, since the free equation has no soliton solutions.

We also assume the existence of some frequency k_{inf}, which has the following property. Outside a certain box $[-L_{inf}, L_{inf}]^N$, the majority of the solution is comprised of gaussians with the property that if $\vec{a}_j x_0 \geq L_{inf}$, then $\vec{b}_j k_0 \geq k_{inf}$ (respectively if $\vec{a}_j x_0 \leq -L_{inf}$, then $\vec{b}_j k_0 \leq -k_{inf}$). This implies that any part of the solution which has moved outside the box $[-L_{inf}, L_{inf}]^N$ is moving outward.

Roughly, what this means is that anything which has already reached the boundary must be moving in the direction of the boundary.

Another difficulty of our method is that it requires a buffer region in which we filter outgoing waves. This buffer region needs to have width $O(k_{inf}^{-1})$, and should encompass many data points (in our examples we typically use approximately 128-512 data points). For comparison, most Dirichlet-to-Neumann based approaches will use far fewer (just enough to numerically calculate a few derivatives). However, those approaches are typically nonlocal in time, and instead need to use many data points in t rather than in x.

Regardless, in both cases, the computational cost on the boundary is orders of magnitude smaller than the computational cost simply to solve the problem on the interior region. See also [63], where the problem of low frequencies is treated with cost proportional to $\log k_{inf}$ rather than k_{inf}^{-1}.

In a recent work [17], Becker, Sewell and Tebbutt show that it is possible to compute the solution on all space in general. In order to do this, one needs to know the solution becomes a free wave far away, that is true for dispersive equations without solitons or other coherent structures moving to infinity. Therefore, our assumptions may be optimal. We would like to mention that in the radial case, the solutions become free outside some compact region [48]. The rigorous analysis carried out in this manuscript has been developed in [65]. The method is also being developed for several types of wave equations in [52].

---

[4] Numerical experiments suggest that our method can also filter outgoing solitons in certain cases, with reasonable accuracy. This is, however, more a coincidence than anything else. It would not occur if one applied this scheme to, e.g. the KdV equation.

1.1.5 Multiscale Extension

In the computations of time-dependent waves, it is expensive to solve a problem in which there is a mixture between low frequencies (shortwaves, with wavenumber $k = k_{sup}$) and high frequencies (longwaves, near $k = k_{inf}$). Let us consider the situation in which low frequencies scatter off a sharp impurity that generates high frequencies propagating and spreading throughout the computational domain, while the domain must be large enough to contain several longwaves. Standard spectral methods have a computational cost being proportional to $O(k_{sup}/k_{inf} \log(k_{sup}/k_{inf}))$. Here, L is the width of the region where the algorithm resolves all frequencies. An advantage of our proposed method is to reduce the cost to $O(k_{sup}L \log(k_{sup}/k_{inf}) \log(k_{sup}L))$ via a multiscale extension. Due to the lack of space, we refer the readers to [63] for a more detailed discussions on this extension.

1.2 Comparison to Other Methods

A variety of other approaches have been proposed for open boundaries. They fall into two main categories, and we discuss them both briefly.

1.2.1 Absorbing Boundary Conditions

High-order absorbing (or artificial) boundary conditions (ABCs) [2, 13, 15, 16, 19, 31–33, 35, 38, 43, 44, 50, 55, 68] are one of the most well-known techniques that allows one to truncate (or approximate) the original equations into a smaller bounded computational domain. Over the last 30 years, different ABCs have been designed for solving the nonlinear Schödinger equations in the literatures [4–9, 11, 13, 14, 24, 40, 46, 47, 73, 75]. We refer to [2, 12] for nice reviews on this topic. The closest approach to ours is the original Engquist-Majda boundary conditions, found in [32, 33]. The principle that was guiding them was that near the boundary, the geometric optics approximation to wave flow is sufficiently accurate to filter off the outgoing waves.

Our result is a direct analogue of this—the gaussian framelet elements behave (under the free flow) like classically free particles. We use a different method to filter, but the guiding principle is the same.

In comparison, the approach that is farthest from ours are the various modern extensions to [32]. Modern approaches attempt to construct the exact solution on the boundary, and then impose it as a boundary condition. In principle, this is the best possible approach. However, in practice, this will be very difficult.

In fact, this approach is sufficiently difficult that we know of few aproaches for the Schrödinger equation. We describe the two main approaches we are aware of, and remark that only the paradifferential strategy of Szeftel attempts to deal with nonlinear equations.

One additional problem with this approach is that it precludes the use of spectral methods to solve the interior problem. The spectrum of $-\Delta$ with Dirichlet-to-Neumann boundaries is not (to our knowledge) known, and even if it were, it is unlikely that an algorithm such as the Fast Fourier Transform would exist for it. This would make spectral methods impractical. For this reason, most implementations of the Dirichlet-to-Neumann approach use FTDT to solve the interior problem, which is another drawback to the Dirichlet-to-Neumann approach.

Exact Dirichlet-Neumann Maps for the Schrödinger Equation

To deal with the free Schrödinger equation (no nonlinearity or potential),Lubich and Schädle [49, 59, 60] as well as Greengard and Jiang [41, 42] constructed methods for using the exact boundary conditions rather than an approximate one. Roughly, their methods consist of approximating the integral kernel by using a piecewise exponential approximation in time and the fact that convolution with an exponential can be done in linear time. This approach appears to work nicely for the free Schrödinger equation, although it is uncertain that it could be applied to the full Dirichlet-to-Neumann operator of a nonlinear equation.

Paradifferential Strategy

In finding the Dirichlet-to-Neumann (DtN) map for the Schrödinger operator without and/or with external potential and nonlinearity at the boundary of the bounded computational domain, using the continuous and/or discrete Laplace transform is the main strategy [5–9, 13, 56]. As the DtN map is a nonlocal time-dependent pseudo-differential operator, its approximations which involve time fractional derivatives and integrals, are commonly used in practical computations. The key of the method is to avoid or to minimize the waves reflected back inside the computational domain while the waves should be outgoing. We refer to [2] for a nice review on this. As constructing the fully nonlinear Dirichlet-to-Neumann operator requires hard-cord analysis, in [66], Szeftel constructs the Dirichlet-to-Neumann operator by a modified version of the paradifferential calculus (introduced in [20]). His methodology is demonstrated in 1 space dimension, with a nonlinearity that is C^∞ in x, $\Psi(\vec{x}, t)$ and $\partial_x \Psi(\vec{x}, t)$. He proves local well posedness of (1.1.1) with his boundary conditions, assuming H^6 regularity of the initial data. However, extensions to $\mathbb{R}^N$ appear highly nontrivial. The assumptions are significantly stronger than ours, and there are no error bounds. However, the numerical experiments are very promising and the results appear accurate for radiative problems (see also Remark 11.3.1).

1.2.2 Absorbing Potentials and Perfectly Matched Layers

Absorbing Potentials

Absorbing (complex) potentials, described in [51], are the current "industry standard". The approach consists of the following. Instead of solving (1.1.1) on the box $[-L_{in}, L_{in}]^N$, we solve

$$i\partial_t \Phi(\vec{x}, t) = -(1/2)\Delta\Phi(\vec{x}, t) + g(t, \vec{x}, \Phi(\vec{x}, t))\Phi(\vec{x}, t) + -ia(x)\Phi(\vec{x}, t)$$

on the region $[-(L_{in} + w), (L_{in} + w)]^N$. The function $a(x)$ is a positive function supported in $[-(L_{in} + w), (L_{in} + w)]^N \setminus [-L_{in}, L_{in}]^N$. The term $-ia(x)$ is a dissipative term which is localized on waves which have left the region $[-L_{in}, L_{in}]^N$. Thus, it (partially) absorbs waves which have left the domain of interest.

This approach is the mainstay of absorbing boundaries, due to it's generality and simplicity. One important reason for the attractiveness is that they are compatible with the FFT/Split Step algorithm (Algorithm 4.3.2), with minimal difficulty of programming.

Of course, the potential $a(x)$ must be tuned to the given problem. Given k_{inf}, k_{sup}, one must select the height and width of the absorber so that it kills most of the wave between k_{inf} and k_{sup}, resulting in an error τ.

Waves with momentum lower than k_{inf} are mostly reflected, and waves with momentum higher than k_{sup} are mostly transmitted (and therefore wrap around the computational domain).

Heuristic calculations and numerical experiments suggest that the absorber must have width proportional to $Ck_{sup}\ln(\epsilon)/k_{inf}$, with C depending on the precise shape of the potential. In contrast, our method works on a boundary layer of width $C\ln(\epsilon)/k_{inf}$, which is smaller by a factor of k_{sup}.

An additional problem with absorbing potentials is that they kill everything on the boundary. They make no distinction between incoming and outgoing waves, and thus they absorb some waves which should return to the region of interest. This poses a fundamental limitation on their use, especially in problems where the nonlinearity creates long range effects, which was illustrated in Sect. 11.2.

Perfectly Matched Layers

Perfectly Matched Layers (PML) are a variation on this approach, proposed by several authors in [2, 3, 10, 37, 53, 76] for the Schrödinger equation (see also [18], where they are first proposed for Maxwell's equations). For comparison of the performance and effectiveness of different PMLs and/or ABCs for nonlinear Schrödinger equations, we refer to [6, 12, 34, 45, 71, 72, 74] and references therein.

To use a PML, instead of solving (1.2.2), we solve:

$$i\partial_t \Phi(\vec{x}, t) = -(1 + ia(x))(1/2)\Delta\Phi(\vec{x}, t) + g(t, \vec{x}, \Phi(\vec{x}, t))\Phi(\vec{x}, t)$$

where $a(x)$ is now a function chosen rather carefully (see below).

The PML has two main advantages over complex absorbing potentials. First, the fact that $ia(x)$ is now in the coefficient of Δ now means that high momentum waves are dissipated more strongly than low momentum ones. Thus, fast waves do not pass through the absorbing potential without being dissipated.

Second, the function $a(x)$ can be chosen precisely so that there is no reflection at the interface (the boundary of $[-L_{in}, L_{in}]^N$). However, this does not eliminate all reflections, as some reflections will be created in the region $[-(L_{in} + w), (L_{in} + w)]^N \setminus [-L_{in}, L_{in}]^N$. The proper way to choose the function $a(x)$ is by exterior complex scaling.

The PML has the same problem as complex absorbing potentials with regards to dissipating incoming waves on the boundary. Additionally, in the presence of a potential, the PML must be modified by applying complex scaling to the potential as well; if the potential is not analytic, this cannot be done. For a nonlinear equation, it actually requires analytically continuing $|\Psi|^2$, which is not a straightforward matter.

Lastly, some PML methods are unstable after discretization. Numerical experiments in [58] suggest that the PML for 2 dimensional Maxwell's Equations exhibit a long time instability. It is possible that this effect occurs for the Schrödinger equation as well.

The PML method for the Schrödinger equation is still very much undeveloped. This makes a more detailed comparison difficult to make.

Chapter 2
Definitions, Notations and A Brief Introduction to Frames

2.1 Definitions and Notations

We will solve (1.1.1) on the region $[-L_{\text{trunc}}, L_{\text{trunc}}]^N$, which is a larger domain than $[-L_{\text{in}}, L_{\text{in}}]^N$. The extra region $[-L_{\text{trunc}}, L_{\text{trunc}}]^N \setminus [-L_{\text{in}}, L_{\text{in}}]^N$ is a buffer region in which we will filter the outgoing waves.

Definition 2.1.1 We define Δ_b to be the Laplacian on the box $[-L_{\text{trunc}}, L_{\text{trunc}}]^N$ with periodic boundary conditions.

Definition 2.1.2 We define $\mathfrak{U}(t)$ to be the propagator of (1.1.1) on $\mathbb{R}^N$. That is, $\mathfrak{U}(t)$ is the map taking $\Psi_0(x) \mapsto \Psi(\vec{x}, t)$ where $\Psi(\vec{x}, t)$ solves (1.1.1) with initial condition $\Psi(\vec{x}, t) = \Psi_0(x)$.

For an initial condition Ψ_0, we define $\mathfrak{U}(t|\Psi_0(x))$ to be the mapping $\Psi_1(x) \mapsto \Psi_1(\vec{x}, t)$ where $\Psi_1(\vec{x}, t)$ solves (2.1.1) with initial condition $\Psi_1(x, 0) = \Psi_1(x)$:

$$\partial_t \Psi_1(\vec{x}, t) = -(1/2)\Delta \Psi_1(\vec{x}, t) + g(t, \vec{x}, \mathfrak{U}(t)\Psi_0)\Psi_1(\vec{x}, t) \qquad (2.1.1)$$

Similarly, $\mathfrak{U}_b(t)$ is the propagator associated to (1.1.1), but with $(1/2)\Delta_b$ replacing $(1/2)\Delta$ and $[-L_{\text{trunc}}, L_{\text{trunc}}]^N$ replacing $\mathbb{R}^N$.

Definition 2.1.3 We make the following conventions regarding notation.

$$|\vec{x}|_p = \left(\sum_{j=1}^{N} |\vec{x}_j|^p \right)^{1/p} \quad \text{for } \vec{x} \in \mathbb{R}^N$$

© The Author(s), under exclusive license to Springer Nature Singapore Pte Ltd. 2023
A. Soffer et al., *Time Dependent Phase Space Filters*, SpringerBriefs on PDEs and Data Science, https://doi.org/10.1007/978-981-19-6818-1_2

We let $d(\vec{x}, \vec{y})$ denote the Euclidean metric on $\mathbb{R}^N$, i.e. $d(\vec{x}, \vec{y}) = |\vec{x} - \vec{y}|_2$. Also, if $A, B \subseteq \mathbb{R}^N$, then:

$$d(\vec{x}, A) = \inf_{\vec{y} \in A} d(\vec{x}, \vec{y})$$

$$d(A, B) = \inf_{\vec{x} \in A, \vec{y} \in B} d(\vec{x}, \vec{y})$$

Definition 2.1.4 We use the notation:

$$\langle x \rangle = (1 + |x|_2^2)^{1/2}$$

We define certain constants related to this notation:

$$\mathfrak{L}_s = \sup_{\vec{x} \in \mathbb{R}^N} \langle x \rangle^s / (1 + |\vec{x}|_s^s)$$

$$\mathfrak{L}_d = \sup_{\vec{x}} \frac{|\nabla \langle \vec{x} \rangle|}{\langle \vec{x} \rangle}$$

Thus:

$$\langle x \rangle^s \leq \mathfrak{L}_s (1 + |\vec{x}|_s^s)$$

Definition 2.1.5 We define the Fourier transform by:

$$\hat{f}(\vec{k}) = (2\pi)^{-N/2} \int_{\mathbb{R}^N} e^{i\vec{k} \cdot \vec{x}} d\vec{x}$$

The inverse Fourier transform is defined by:

$$f(\vec{x}) = (2\pi)^{-N/2} \int_{\mathbb{R}^N} e^{-i\vec{k} \cdot \vec{x}} d\vec{k}$$

Thus, the operator $f(\vec{x}) \mapsto \hat{f}(\vec{k})$ is an isometry from $L^2(\mathbb{R}^N, d\vec{x}) \to L^2(\mathbb{R}^N, d\vec{k})$, and $\|f(\vec{x})\|_{L^2(\mathbb{R}^N, d\vec{x})} = \left\|\hat{f}(\vec{k})\right\|_{L^2(\mathbb{R}^N, d\vec{k})}$.

Definition 2.1.6 We define the Sobolev spaces $H^s = H^s(\mathbb{R}^N)$ by the norms:

$$\|f\|_{H^s}^2 = \|f\|_{L^2(\mathbb{R}^N)}^2 + \sum_{j=1}^{N} \left\|\partial_{x_j}^s f\right\|_{L^2(\mathbb{R}^N)}^2 \tag{2.1.2}$$

We make this particular choice of definition when we compute the constants. Similarly, we define the Sobolev spaces H_b^s by the norms:

$$\|f\|_{H_b^s}^2 = \|f\|_{L^2([-L_{\text{trunc}}, L_{\text{trunc}}]^N)}^2 + \sum_{j=1}^{N} \left\| \partial_{x_j}^s f \right\|_{L^2([-L_{\text{trunc}}, L_{\text{trunc}}]^N)}^2$$

We define the constant

$$\mathbf{h}_s^{\pm} = \sup_{f \in H^s} \left(\|f\|_{H^s} / \left\| \langle \vec{k} \rangle^s \hat{f}(\vec{k}) \right\|_{L^2} \right)^{\pm 1} = \sup_{\vec{k} \in H^s} \left((1 + |\vec{k}|_s^s) / \langle \vec{k} \rangle^s \right)^{\pm 1}$$

This allows us to relate the Sobolev space we use to Sobolev spaces defined by using $\langle \vec{k} \rangle$.

No matter which Sobolev space we work in, we always let $\langle \cdot | \cdot \rangle$ denote the inner product in L^2.

Definition 2.1.7 We make use of smoothed out characteristic functions. Let A be a closed set and let w be a positive number. Toward that end, we demand that the function $P_{A;w}^s(\vec{x})$ have the following properties:

1. $P_{A;w}^s(\vec{x}) = 1$ for $\vec{x} \in A$, and $P_{A;w}^s(\vec{x}) = 0$ if the euclidean distance between $\vec{x}$ and A is greater than w.
2. $\partial_{x_j}^k P_{A;w}^s(\vec{x})$ exists and is continuous for $j = 1..N, k = 1..s$.
3. $P_{A;w}^s(\vec{x})$ has minimal norm as an operator from $H^s \to H^s$.

We adopt the convention that $P_{A;w}^0(\vec{x}) = 1_A(\vec{x})$, that is, $P_{A;w}^0(\vec{x}) = 0$ for $\vec{x} \notin A$ regardless of w.

Definition 2.1.8 We define $\mathbf{m}_{c,s}(\sigma, N)$, $\mathbf{m}_{v,s}(\sigma, N)$ and $\mathbf{m}_{c,s}'(\sigma, N)$, $\mathbf{m}_{v,s}'(\sigma, N)$ so that

$$\int_{\mathbb{R}^N} \langle \vec{x} \rangle^s e^{-|\vec{y}-\vec{x}|/\sigma^2} d\vec{y} \leq \mathbf{m}_{c,s}(\sigma, N) + \mathbf{m}_{v,s}(\sigma, N) |\vec{x}|_2^s$$

$$\int_{\mathbb{R}^N} \langle \vec{x} \rangle^s \left| \nabla e^{-|\vec{y}-\vec{x}|/\sigma^2} \right| d\vec{y} \leq \mathbf{m}_{c,s}'(\sigma, N) + \mathbf{m}_{v,s}'(\sigma, N) |\vec{x}|_2^s$$

2.2 A Brief Introduction to Frames

We first discuss briefly the concept of a frame, which will be crucial to our analysis. A frame is basically an overcomplete basis for a Hilbert space, in our case, $L^2(\mathbb{R}^N)$. A framelet decomposition is the tool we use to break up the solution $\Psi(\vec{x}, t)$ into incoming and outgoing components.

Definition 2.2.1 A frame is a countable set of functions (in some Hilbert space, e.g. L^2) $\{\phi_j(x)\}_{j \in J}$ (for some index set J) such that $\exists \mathcal{A}_F, \mathcal{B}_F$ such that for any $f \in L^2(\mathbb{R}^N)$:

$$\mathcal{A}_F \|f\|_{L^2} \leq \left\| \langle f(x) | \phi_j(x) \rangle \right\|_{l^2(J)} \leq \mathcal{B}_F \|f\|_{L^2}$$

The framelet analysis operator F is the map $f(x) \mapsto \vec{f} \in l^2(J)$, where $\vec{f}_j = \langle f | \phi_j(x) \rangle$.

The individual vectors $\phi_j(x)$ are referred to as framelets, and $j \in J$ are referred to as framelet indices.

Definition 2.2.2 For a frame $\{\phi_j(x)\}_{j \in J}$, the dual frame $\left\{ \tilde{\phi}_j(x) \right\}_{j \in J}$ is the unique frame such that:

$$\tilde{\phi}_j(x) = (F^\star F)^{-1} \phi_j(x)$$

where $F^\star : l^2(J) \to L^2(\mathbb{R}^N)$ is the adjoint of F. It is the "best" (see below for an explanation) set of vectors such that for all $f(x) \in L^2$:

$$f(\vec{x}) = \sum_{j \in J} \left\langle \tilde{\phi}_j(x) | f(x) \right\rangle \phi_j(x)$$

The dual frame is also a frame, with frame bounds $\mathcal{B}_F^{-1}$ and $\mathcal{A}_F^{-1}$.

The framelet coefficients of a function $f(x)$, are the "best" set of coefficients such that:

$$f(x) = \sum_{j \in J} f_j \phi_j(x)$$

The framelet coefficients are not unique. By "best", we mean that $\vec{f}_j$ is the collection of framelet indices minimizing

$$\sum_{j \in J} |f_j|^2 .$$

They can be calculated by the formula:

$$f_{(\vec{a}, \vec{b})} = \left\langle \tilde{\phi}_j(x) | f(\vec{x}) \right\rangle \tag{2.2.1}$$

For a function $f(x, t)$ depending on time, we denote by $f_j(t)$ the framelet coefficients of $f(\vec{x}, t)$ at time t.

2.2.1 Windowed Fourier Transform

As an example, we can let $J = \mathbb{Z}^N \times \mathbb{Z}^N$ and let the individual framelets $\phi_{(\vec{a},\vec{b})}(\vec{x})$ be given by:

$$\phi_{(\vec{a},\vec{b})}(\vec{x}) = \pi^{-N/4}\sigma^{-N/2}e^{ik_0\vec{b}\cdot\vec{x}}e^{-|\vec{x}-\vec{a}x_0|_2^2/2\sigma^2}$$

For $\sigma \in \mathbb{R}^+$ and $x_0, k_0 \in \mathbb{R}^+$ such that $x_0k_0 \leq 2\pi$, then the set

$$\left\{\phi_{(\vec{a},\vec{b})}(\vec{x})\right\}_{(\vec{a},\vec{b})\in\mathbb{Z}^N\times\mathbb{Z}^N}$$

is a frame in $L^2(\mathbb{R}^N)$. This is known as the windowed Fourier transform frame (with Gaussian window), abbreviated WFT frame. We will return to this specific example later, in Sect. 3. This is the frame we use to build the outgoing wave filter.

Subject to additional conditions on x_0, k_0 and σ, the WFT can also form a frame in various Sobolev spaces (see Theorem 3.1.5, proved in [27], and Corollary 3.1.6).

2.2.2 Localization of Phase Space

For the WFT filter, we consider the index set $\mathbb{Z}^N \times \mathbb{Z}^N$ to be a discrete representation of phase space. That is, we consider the point $(\vec{a}, \vec{b})$ to represent the point $(\vec{a}x_0, \vec{b}k_0)$ in phase space.

For a frame that is well localized in phase space, it is simple to characterize the flow with respect to $e^{i(1/2)\Delta t}$. Under the free flow, individual framelets behave like classical particles. For instance, the Gaussian framelet $\phi_{(\vec{a},\vec{b})}(\vec{x})$ travels along the trajectory $\vec{a}x_0 + t\vec{b}k_0$ when propagated by $e^{i(1/2)\Delta t}$. Due to the heisenberg uncertainty principle, the framelet also spreads out at the rate t/σ. When $\vec{b}k_0 \gg \sigma$, it is simple to determine whether the framelet is moving inward or outward, and delete it as is necessary. Of course, of $\vec{b}k_0$ is very close to zero, then the spreading will be the dominant mode of transport. This is the largest source of error in our method.

Some other frames also provide good localization in phase space, although in different ways. For instance, frames of wavelets travel consistently along classical trajectories, but with the added cost that more slowly moving framelets are spread out more in space (as opposed to the Gaussian WFT, for which all framelets have the same width).

It appears very likely that one could replace the WFT frame that we use by a frame of wavelets, or other frames, provided they have the appropriate phase space localization properties.

In addition, we remark on one extremely promising possibility for extending our analysis to hyperbolic systems. It was proved recently by Demanet and Candes (c.f.

[23]) that a curvelet frame allows for a sparse representation of wave propagators in the high frequency regime. We intend to investigate the possibility of using curvelets to construct a boundary filter for dispersive hyperbolic systems, e.g. Maxwell's equations.

2.2.3 Distinguished Sets of Framelets, Framelet Functionals

We now define certain distinguished sets of framelets, and also two relevant framelet functionals. Namely, we define the per-framelet error, and per-framelet relevance functions. The per framelet error functional is a measure of the difference between the propagators $e^{i(1/2)\Delta t}$ and $\mathfrak{U}(t)$ when applied to that particular framelet. Similarly, the per-framelet relevance functional is a measure of how important a particular framelet is to the solution inside the box.

Definition 2.2.3 For a frame $\{\phi_j\}$, a Sobolev space H^s and a distance L_{in} (to be specified later), we define a family of functions, the relevance functions to be:

$$\left\| e^{i(1/2)\Delta t} \phi_j \right\|_{H^s([-L_{\text{in}}, L_{\text{in}}]^N)} = \mathcal{R}_j^s(t) \tag{2.2.2}$$

We now define the set of bad framelets, that is, those framelets which cause most of the short time error. Ideally, these are the ones we would like to filter (although this will not be possible).

Definition 2.2.4 For a frame $\{\phi_j\}$ and a Sobolev space H_b^s, we define a family of functions, the per-framelet error functions to be a set of functions $\mathfrak{E}_j^s(t)$ such that:

$$\left\| (e^{i(1/2)\Delta t} - e^{i(1/2)\Delta_b t}) \phi_j \right\|_{[-(L_{\text{in}}+w), (L_{\text{in}}+w)]^N} \leqslant \mathfrak{E}_j^s(T) \tag{2.2.3}$$

These will be computed for the WFT frame later on.

Definition 2.2.5 For a frame $\{\phi_j\}$, a Sobolev space H_b^s an error tolerance ε, and a time T (possibly ∞), we define the set of error causing framelets BAD(ε, s, T) to be:

$$\text{BAD}(\varepsilon, s, T) = \{ j \in J | \exists t < T \text{ such that } \mathfrak{E}_j^s(t) > \varepsilon \} \tag{2.2.4}$$

Definition 2.2.6 The Big Box is defined by:

$$\text{BB}(\delta_{\text{BB}})$$

$$= ([-(L_{\text{in}}+w+\mathfrak{X}_{\square}^s(\epsilon, k_{\text{sup}}, L_{\text{in}}+w)), (L_{\text{in}}+w+\mathfrak{X}_{\square}^s(\epsilon, k_{\text{sup}}, L_{\text{in}}+w))]^N \cap x_0 \mathbb{Z}^N)$$

$$\times ([-k_{\text{sup}}, k_{\text{sup}}]^N \cap k_0 \mathbb{Z}^N)$$

We define the computational width, L_{trunc}, by:

$$L_{\text{trunc}} = L_{\text{in}} + w + \mathfrak{X}_{\square}^s(\epsilon, k_{\text{sup}}, L_{\text{in}} + w)$$

The number $\mathfrak{X}_{\square}^s(\epsilon, k_{\text{sup}}, L_{\text{in}} + w)$ is an extra buffer region needed due to the widthe of the framelets. We define it precisely.

Definition 2.2.7 Let $B_X = [-X, X]^N$, $B_K = [-K, K]^N$ for $X, K < \infty$. Then $\mathfrak{X}_{\square}^s(\epsilon, K, X)$ and $\mathfrak{K}_{\square}^s(\epsilon, K)X$ are the smallest numbers for which the following estimate holds.

Let $X' = X - \mathfrak{X}_{\square}^s(\epsilon, K, X)$, $K' = K - \mathfrak{K}_{\square}^s(\epsilon, K)$. Then:

$$\left\| f(x) - \mathcal{P}_{B_{X'} \times B_{K'}} f(x) \right\|_{H^s} \leq \mathfrak{H}_+^s(\tilde{g}(\vec{x})) \mathfrak{H}_+^{-s}(e^{-x^2/\sigma^2})$$
$$\times \left(\left\| (1 - P_{B_X;x_0}^s(\vec{x})) f(\vec{x}) \right\|_{H^s} + \left\| (1 - P_{B_K;k_0}^0(\vec{k})) f(\vec{x}) \right\|_{H^s} + \epsilon \, \| f \|_{H^s} \right)$$

$$(2.2.5)$$

We provide a proof that this definition is not vacuous in Theorem 3.3.7.

We note that when we solve (1.1.1) with periodic boundary conditions, we will do so on the box $[-L_{\text{trunc}}, L_{\text{trunc}}]^N$.

The set $\text{NECC}(\varepsilon, s, t)$ is the set of framelets which have a nontrivial incoming component. That is, these are the framelets which will return to the region of interest, at least partially. $\text{NECC}(\varepsilon, s, T)$ should be thought of as "incoming waves", and cannot be filtered without causing error.

Definition 2.2.8 For a frame $\{\phi_j\}$, a Sobolev space H_b^s an error tolerance ε, and a time T (possibly ∞), we define the set $\text{NECC}(\varepsilon, s, T)$ to be:

$$\text{NECC}(\varepsilon, s, T) = \{ j \in J | \exists t < T \text{ such that } \mathcal{R}_j^s(t) > \varepsilon \} \qquad (2.2.6)$$

Chapter 3
Windowed Fourier Transforms and Space Phase Numerics

In this part, we review some basic results on frames and the windowed Fourier transform. More detailed information can be found in [27, 28, 30], for example.

3.1 Basic Definitions and Properties

The discrete windowed Fourier transform frame is the standard frame of canonical coherent states. We use it because of it's excellent time and frequency localization properties (if a Gaussian window is used), as well as the fact that we have strong, rigorous control on the properties which we use. In practice, other frames can be used if they are more effective for some particular problem.

Definition 3.1.1 The Gaussian WFT frame is the set of functions

$$\left\{ \phi_{(\vec{a},\vec{b})}(\vec{x}) = \pi^{-N/4} \sigma^{-N/2} e^{ik_0 \vec{b}\cdot\vec{x}} e^{-|\vec{x}-\vec{a}x_0|_2^2/2\sigma^2} \right\}_{(\vec{a},\vec{b}) \in \mathbb{Z}^N \times \mathbb{Z}^N}$$

for some x_0, k_0, σ. To be a frame, $x_0 k_0 < 2\pi$, otherwise there exist vectors orthogonal to the span of the WFT frame. The dual frame to the Gaussian WFT frame is also a WFT frame, given by

$$\left\{ e^{ik_0 \vec{b}\cdot\vec{x}} \tilde{g}(\vec{x} - \vec{a}x_0) \right\}_{(\vec{a},\vec{b}) \in \mathbb{Z}^N \times \mathbb{Z}^N}$$

for a certain $\tilde{g} \in L^2(\mathbb{R}^N)$ (clarified later).

We will refer to $\phi_{(\vec{a},\vec{b})}(\vec{x})$ as a framelet localized at $(\vec{a}x_0, \vec{b}k_0)$ in phase space. When we refer to the position or velocity of a framelet, we are referring to $\vec{a}x_0$ and $\vec{b}k_0$, respectively.

© The Author(s), under exclusive license to Springer Nature Singapore Pte Ltd. 2023
A. Soffer et al., *Time Dependent Phase Space Filters*, SpringerBriefs on PDEs and Data Science, https://doi.org/10.1007/978-981-19-6818-1_3

The following theorem establishes that the WFT is a frame, in the special case when $x_0 k_0 = 2\pi/\mathfrak{M}$, for some $\mathfrak{M} \in \mathbb{N}$. The number $\mathfrak{M}$ is called the oversampling rate. It also explicitly provides the frame bounds.

We remark here that throughout this paper, we will always take $x_0 k_0 = 2\pi/\mathfrak{M}$, with $\mathfrak{M}$ an even integer. We do this in order to use both Theorem 3.1.4 and also Theorem 3.2.2 (which is stated later).

We conjecture that a similar result holds for $\mathfrak{M} \in (1, \infty)$. The assumption $\mathfrak{M} \in 2\mathbb{Z}$ is made for algebraic simplicity, and very likely is unnecessary.

Before stating our results, we introduce a transform, the Zak transform, which is useful in the analysis of the WFT. In a certain sense, the Zak transform is used to "diagonalize" the WFT; this is explained in more detail in the proof of Theorem 3.1.4.

Definition 3.1.2 The Zak transform is the isometry $\mathbf{Z} : L^2(\mathbb{R}^N) \to L^2([0, 1]^N \times [0, 1]^N)$ defined by:

$$(\mathbf{Z}f)(\vec{t}, \vec{s}) = x_0^{N/2} \sum_{\vec{l} \in \mathbb{Z}^N} e^{2\pi i(\vec{t} \cdot \vec{l})} f(x_0(\vec{s} - \vec{l})) \tag{3.1.1}$$

and for $\phi(\vec{t}, \vec{s}) \in L^2([0, 1]^N \times [0, 1]^N)$:

$$\mathbf{Z}^{-1}\varphi(\vec{x}) = x_0^{N/2} \int_{[0,1]^N} e^{-2\pi i(\vec{t} \cdot \lfloor \vec{x}/x_0 \rfloor)} \phi\left(\vec{t}, \vec{x}/x_0\right) dt \tag{3.1.2}$$

Note that $(\mathbf{Z}f)(\vec{t}, \vec{s})$ is 1-periodic $\vec{t}$.

We also review some results concerning the $\theta_3(z|\tau)$, which we use shortly.

Definition 3.1.3 The function $\theta_3(z|\tau)$ is defined by:

$$\theta_3(z|\tau) = 1 + 2\sum_{l=1}^{\infty} \cos(2\pi l z)e^{i\pi \tau l^2} \tag{3.1.3}$$

It has the equivalent definition:

$$\theta_3(z|\tau) = \prod_{n=1}^{\infty} (1 - e^{i2\pi n\tau})(1 + e^{(2n-1)i\pi\tau}e^{2\pi iz})(1 + e^{(2n-1)i\pi\tau}e^{-2\pi iz}) \tag{3.1.4}$$

It can be analytically continued in z by the recurrence relation:

$$\theta_3(z + \tau, \tau) = e^{-\pi i(\tau - 2z)}\theta_3(z, \tau) \tag{3.1.5}$$

We now use the Zak transform to compute frame bounds.

Theorem 3.1.4 (See [29]) *Let F be the framelet analysis operator for a windowed Fourier transform, i.e. the mapping* $F : L^2(\mathbb{R}^N) \to l^2(\mathbb{Z}^N \times \mathbb{Z}^N)$ *given by* $[F\Psi(x)]_{\vec{a},\vec{b}} = \langle \phi_{(\vec{a},\vec{b})}(\vec{x}) | \Psi(x) \rangle$. *Suppose that for some integer* $\mathfrak{M} \geq 2$, $x_0 k_0 = 2\pi/\mathfrak{M}$. *Define:*

$$S(x_0, \mathfrak{M}, \vec{t}, \vec{s}) = \left| [\mathbf{Z} e^{-x^2/2}](\vec{s}, \vec{t}) \right|^2$$

$$= \left(\frac{x_0}{\sqrt{\pi}} \right)^N \sum_{\vec{r} \in \{0, \dots \mathfrak{M}-1\}^N} \left| \sum_{\vec{l} \in \mathbb{Z}^N} \exp\left(2\pi i \vec{l} \cdot (\vec{t} - \vec{r}/\mathfrak{M}) \right) \exp\left(\frac{-x_0^2}{2}(\vec{s} - \vec{l})^2 \right) \right|^2$$

$$= \frac{x_0^N e^{-|s|^2 x_0^2}}{\pi^{N/2}} \sum_{\vec{r} \in \{0, \dots \mathfrak{M}-1\}^N} \prod_{j=1}^N \theta_3 \left(\vec{t}_j \frac{\vec{r}_j}{\mathfrak{M}} + i \frac{x_0^2}{2\pi} \vec{s}_j \left| \frac{i x_0^2}{2\pi} \right. \right) \times$$

$$\theta_3 \left(\vec{t}_j - \frac{\vec{r}_j}{\mathfrak{M}} - i \frac{x_0^2}{2\pi} \vec{s}_j \left| \frac{i x_0^2}{2\pi} \right. \right) \qquad (3.1.6)$$

Then:

$$[\mathbf{Z} F^* F \mathbf{Z}^{-1} f](\vec{t}, \vec{s}) = S(x_0, \mathfrak{M}, \vec{t}, \vec{s}) f(\vec{t}, \vec{s}) \qquad (3.1.7)$$

This implies that:

$$\mathcal{A}_F = \inf_{(\vec{s}, \vec{t}) \in [0,1]^{N+N}} \left| S(x_0, \mathfrak{M}, \vec{t}, \vec{s}) \right| \qquad (3.1.8a)$$

$$\mathcal{B}_F = \sup_{(\vec{s}, \vec{t}) \in [0,1]^{N+N}} \left| S(x_0, \mathfrak{M}, \vec{t}, \vec{s}) \right| \qquad (3.1.8b)$$

Proof This is proved in [29] for the one dimensional case, where $\sigma = 1$. The multidimensional follows by noting that:

$$S(x_0, \mathfrak{M}, \vec{t}, \vec{s}) = \prod_{j=1}^N S_{1d}(x_0, \mathfrak{M}, \vec{t}_j, \vec{s}_j)$$

The case $\sigma \neq 1$ is recovered by scaling. $\qquad\qquad\qquad\qquad\qquad\qquad \square$

The next theorem is taken from [27]. It shows that for a sufficiently oversampled frame, the WFT is a frame in Sobolev spaces as well.

Theorem 3.1.5 (Daubechies [27]) *Recall the operator:*

$$F^* F f(x) = \sum_{(\vec{a}, \vec{b}) \in \mathbb{Z} \times \mathbb{Z}} e^{ibk_0} g(x - ax_0) \langle e^{ibk_0} g(x - ax_0) | f(x) \rangle$$

where $g(x)$ is either $e^{-x^2/2}$ or the 1 dimensional dual window $\tilde{g}(x)$. The operator F^*F is bounded above and below, in H^s and H^{-s}, provided the constants $A_s(g)$ and $B_s(g)$ (defined below) are strictly positive. This implies that if $A_s(g)$ and $B_s(g)$ are strictly positive, then the GWFT is a frame in $H^s(\mathbb{R})$ and $H^{-s}(\mathbb{R})$.

$$A_s(g) \left\| \langle \partial_x \rangle^{\pm s} f(x) \right\|_{L^2} \leq \left\| \langle \partial_x \rangle^{\pm s} F^* F f(x) \right\|_{L^2} \leq B_s(g) \left\| \langle \partial_x \rangle^{\pm s} f(x) \right\|_{L^2}$$

We must first construct some auxiliary functions. Define:

$$m(\hat{g}; k_0) = \inf_{x \in \mathbb{R}} \sum_{b \in \mathbb{Z}} \left| \hat{g}(k + bk_0) \right|^2 \tag{3.1.9a}$$

$$M(\hat{g}; k_0) = \sup_{x \in \mathbb{R}} \sum_{b \in \mathbb{Z}} \left| \hat{g}(k + bk_0) \right|^2 \tag{3.1.9b}$$

Define, for $s \geq 0$:

$$\beta_s^{\pm}(k') = \sup_k \left[\langle k \rangle^{\mp s} \langle k + k' \rangle^{\pm s} \sum_{b \in \mathbb{Z}} \left| \hat{g}(k + bk_0) \right| \left| \hat{g}(k + bk_0 + k') \right| \right]$$

$$A_s(g) = \frac{2\pi}{x_0} \left[m(\hat{g}; k_0) - \sum_{a \neq 0} \left(\beta_s^+(2\pi a/x_0) \beta_s^-(-2\pi a/x_0) \right)^{1/2} \right] \tag{3.1.10a}$$

$$B_s(g) = \frac{2\pi}{x_0} \left[m(\hat{g}; k_0) + \sum_{a \neq 0} \left(\beta_s^+(2\pi a/x_0) \beta_s^-(-2\pi a/x_0) \right)^{1/2} \right] \tag{3.1.10b}$$

Corollary 3.1.6 *In N dimensions, we find that*

$$\mathfrak{H}_-^s(g) \left\| f(\vec{x}) \right\|_{H^{\pm s}} \leq \left\| F^* F f(\vec{x}) \right\|_{H^{\pm s}} \leq \mathfrak{H}_+^s(g) \left\| f(\vec{x}) \right\|_{H^{\pm s}}$$

where

$$\mathfrak{H}_+^s(g) = N B_0(g)^{N-1} B_s(g) \tag{3.1.11a}$$

$$\mathfrak{H}_-^s(g) = N A_0(g)^{N-1} A_s(g) \tag{3.1.11b}$$

Thus, in $H^s(\mathbb{R}^N)$ and $H^{-s}(\mathbb{R}^N)$, the WFT is a frame with frame bounds $\mathfrak{H}_-^s(g)$ and $\mathfrak{H}_+^s(g)$, provided they are both positive.

Table 3.1 Frame Bounds, as
a function of s, for a
particular GWFT frame. The
parameters are $\sigma = 1$,
$x_0 = 1$, $k_0 = \pi/2$. For $s = 7$,
the estimates break down.
This table is taken from [27],
where it is table VI-A

s	A_s	B_s	B_s/A_s
0	3.853	4.147	1.076
1	3.852	4.148	1.077
2	3.849	4.151	1.079
3	3.836	4.164	1.086
4	3.787	4.213	1.112
5	3.600	4.400	1.222
6	2.865	5.135	1.793

Proof We want to compute upper and lower bounds on:

$$\left\| F^* F g(\vec{x}) \right\|_{H^s} = \sum_{j=1}^{N} \left\| (1 + (i\partial_{x_j})^s) F^* F g(\vec{x}) \right\|_{L^2}$$

To the j'th term of the sum, we apply Theorem 3.1.5 in the $j'th$ direction. This pulls out a factor of $A_s(g)$. In the directions $1 \ldots j - 1$ and $j + 1 \ldots N$, we do the same thing, which pulls out a factor of $A_0(g)$ (since there are no derivatives in that direction). We then add up over $j = 1 \ldots N$. Thus we obtain the lower bound. The upper bound is done identically. $\square$

Remark 3.1.7 As one can see from Table 3.1, even for a frame which is oversampled only by $\mathfrak{M} = 4$, the WFT is a reasonably tight frame even in H^3, where it differs from being tight by less than 10%. In practice, for filtering outgoing waves, we will often want a higher oversampling rate to ensure good decay of the dual window, so we expect this will not usually pose a problem.

In fact, we believe this bound is suboptimal, and conjecture that the WFT is a frame in any Sobolev space. But we do not know how to prove it, although the result can probably be tightened using the Zak transform.

We make another observation, about the Sobolev norms of framelets.

Definition 3.1.8 We denote the per-framelet energy by:

$$(\mathcal{M}^s_{(\vec{a},\vec{b})})^2 = \sum_{k=1}^{N} \left\| \partial^s_{x_j} \phi_{(\vec{a},\vec{b})}(\vec{x}) \right\|^2_{L^2(\mathbb{R}^N)} \tag{3.1.12}$$

Also, $\mathcal{M}^0_{(\vec{a},\vec{b})} = 1$.

Note that $\mathcal{M}^0_{(\vec{a},\vec{b})} = 1$. We have the relation $\left\| \phi_{(\vec{a},\vec{b})}(\vec{x}) \right\|^2_{H^s} = (\mathcal{M}^0_{(\vec{a},\vec{b})})^2 + (\mathcal{M}^s_{(\vec{a},\vec{b})})^2 = 1 + (\mathcal{M}^s_{(\vec{a},\vec{b})})^2$.

Proposition 3.1.9 *The framelet energy is bounded by:*

$$\mathcal{M}^s_{(\vec{a},\vec{b})} \leq \mathfrak{F}_s \left(\sum_{k=1}^{N} (2\sigma)^{-s} (\exp_s(\sqrt{2\sigma}\vec{b}_k k_0))^2 \right)^{1/2} \tag{3.1.13}$$

$$\mathfrak{F}_s = \frac{s!}{\sqrt{2\pi}} \left(\int_0^{2\pi} e^{-2\cos(\tau)} d\tau \right)^{1/2} \tag{3.1.14}$$

The function $\exp_s(z)$ *is defined by:*

$$\exp_s(z) = \sum_{j=0}^{s} \frac{z^j}{j!} \tag{3.1.15}$$

Thus, $(\mathcal{M}^s_{(\vec{a},\vec{b})})^2 \leq (\mathfrak{F}_s/s!) \left|\vec{b}k_0\right|^s_s + O\left(\left|\vec{b}k_0\right|^{s-1}_s\right)$ *as* $\left|\vec{b}k_0\right| \to \infty.$

Proof We begin by computing in 1 dimension. We neglect the space translations, which will not effect the mass.

$$\partial_x^s e^{ibk_0x} e^{-x^2/2\sigma^2} = \sum_{j=0}^{s} \binom{s}{j} (ibk_0)^j e^{ibk_0x} \partial_x^{s-j} e^{-x^2/2\sigma^2}$$

$$= e^{ibk_0x} \sum_{j=0}^{s} \binom{s}{j} (ibk_0)^j (-2\sigma)^{-(s-j)/2} H_{s-j}(x/\sqrt{2}\sigma) e^{-x^2/2\sigma^2} \tag{3.1.16}$$

We use the contour integral representation of $H_n(z)$ to write:

$$(3.1.16) = e^{ibk_0x} \sum_{j=0}^{s} \binom{s}{j} (ibk_0)^j (-2\sigma)^{-(s-j)/2} (s-j)!$$

$$\times \int_{|z|=1} e^{-(x/\sqrt{2}\sigma - z)^2} z^{-(s-j)-1} \frac{dz}{2\pi i z}$$

$$= e^{ibk_0x} s! (-2\sigma)^{-s/2} \int_{|z|=1} \exp_s(-\sqrt{2}\sigma bk_0 z) e^{-(x/\sqrt{2}\sigma - z)^2} z^{-(s-1)} \frac{dz}{2\pi i z} \tag{3.1.17}$$

We multiply this by its complex conjugate, and integrate with respect to x:

$$\int \left[(s!)^2 (2\sigma)^{-s} \int_{|z|=1} \int_{|t|=1} \exp_s(-\sqrt{2}\sigma bk_0 z) \exp_s(-\sqrt{2}\sigma bk_0 t) \right.$$

$$\left. e^{-(x/\sqrt{2}\sigma - z)^2} e^{-(x/\sqrt{2}\sigma - t)^2} z^{-(s-1)} \frac{dz}{2\pi i z} t^{-(s-1)} \frac{dt}{2\pi i t} \right] dx$$

$$
= \int \left[(s!)^2 (2\sigma)^{-s} \int_{|z|=1} \int_{|t|=1} \exp_s(-\sqrt{2\sigma} bk_0 z) \exp_s(-\sqrt{2\sigma} bk_0 t) \right.
$$

$$
\left. e^{-(x/\sigma-(t+z))^2} e^{-2tz} z^{-(s-1)} \frac{dz}{2\pi i z} t^{-(s-1)} \frac{dt}{2\pi i t} \right] dx
$$

$$
= \int_{|z|=1} \int_{|t|=1} (s!)^2 (2\sigma)^{-s} \exp_s(-\sqrt{2\sigma} bk_0 z) \exp_s(-\sqrt{2\sigma} bk_0 t) e^{-2tz} z^{-(s-1)}
$$

$$
\times \left(\int e^{-(x/\sigma-(t+z))^2} dx \right) \frac{dz}{2\pi i z} t^{-(s-1)} \frac{dt}{2\pi i t} \tag{3.1.18}
$$

The integral in x is independent of the values of t and z. Thus:

$$
(3.1.18) = \left(\int e^{-x^2/\sigma^2} dx \right) (s!)^2 (2\sigma)^{-s} \int_{|z|=1} \int_{|t|=1}
$$

$$
\exp_s(-\sqrt{2\sigma} bk_0 z) \exp_s(-\sqrt{2\sigma} bk_0 t) e^{-2tz} z^{-(s-1)} \frac{dz}{2\pi i z} t^{-(s-1)} \frac{dt}{2\pi i t} \tag{3.1.19}
$$

We bound the integral by the $L^1 - L^\infty$ duality, to obtain:

$$
|(3.1.19)| = \int (\partial_x^s e^{ibk_0 x} e^{-x^2/2\sigma^2})(\partial_x^s e^{-ibk_0 x} e^{-x^2/2\sigma^2}) dx
$$

$$
\leq \left\| \exp_s(-\sqrt{2\sigma} bk_0 z) \exp_s(-\sqrt{2\sigma} bk_0 t) \right\|_{L^\infty(ds/2\pi i s, dt/2\pi i t)}
$$

$$
\times \left\| e^{-2tz} z^{-(s-1)} t^{-s-1} \right\|_{L^1(ds/2\pi i s, dt/2\pi i t)}
$$

$$
\leq (s!)^2 (2\sigma)^{-s} (\exp_s(\sqrt{2\sigma} bk_0))^2 \left((2\pi)^{-1} \int_0^{2\pi} e^{-2\cos(\theta)} d\theta \right) \tag{3.1.20}
$$

We moved from the second line to the third by computing:

$$
\int_{|z|=1} \int_{|t|=1} \left| e^{-2tz} z^{-s-1} t^{-s-1} \right| \frac{dt}{2\pi t} \frac{dz}{2\pi t}
$$

$$
= (2\pi)^{-2} \int_0^{2\pi} \int_0^{2\pi} e^{-2\cos(\theta-\phi)} d\theta d\phi = (2\pi)^{-2} \int_0^{2\pi} \int_0^{2\pi} e^{-2\cos(\tau)} d\tau d\beta
$$

$$
= (2\pi)^{-1} \int_0^{2\pi} e^{-2\cos(\tau)} d\tau
$$

To finish, we compute:

$$(\mathcal{M}^s_{(\vec{a},\vec{b})})^2 = \sum_{k=1}^{N} \left\| \partial^s_{x_j} \phi_{(\vec{a},\vec{b})}(\vec{x}) \right\|^2_{L^2(\mathbb{R}^N)}$$

$$= \sum_{k=1}^{N} \int_{\mathbb{R}^N} \left| \partial^s_{x_j} \pi^{-N/4} \sigma^{-N/2} e^{ik_0 \vec{b}\cdot\vec{x}} e^{-|\vec{x}-\vec{a}x_0|^2_2/2\sigma^2} \right|^2 d\vec{x}$$

$$\leq \sum_{k=1}^{N} (s!)^2 (2\sigma)^{-s} \left((2\pi)^{-1} \int_0^{2\pi} e^{-2\cos(\tau)} d\tau \right) (\exp_s(\sqrt{2}\sigma \vec{b}_k k_0))^2 \qquad (3.1.21)$$

This is what we wanted to prove.   □

In this subsection, we describe some properties of the WFT frame that we use at various points.

3.2 Characterizing the Dual Window

We now characterize the dual window. Recall that the dual window is the unique function such that

$$f(\vec{x}) = \sum_{(\vec{a},\vec{b}) \in \mathbb{Z}^N \times \mathbb{Z}^N} \left\langle f(x) | e^{ik_0\vec{b}\cdot\vec{x}} \tilde{g}(\vec{x} - \vec{a}x_0) \right\rangle \pi^{-N/4} \sigma^{-N/2} e^{ik_0\vec{b}\cdot\vec{x}} e^{-|\vec{x}-\vec{a}x_0|^2_2/2\sigma^2}$$

for $f(\vec{x}) \in L^2(\mathbb{R}^N)$.

We show that the dual window is exponentially localized in position and momentum, and calculate the constants explicitly (this is Theorem 3.2.2). Our results only apply when $\mathfrak{M} \in 2\mathbb{N}$, but this is merely because the algebra becomes simple in this case. We believe it is likely that similar results will hold for $\mathfrak{M} \notin 2\mathbb{N}$.

Our result implies that as $\mathfrak{M} \to \infty$, the exponential decay rate of $\tilde{g}(x)$ grows without bound. This is to be expected, since the dual window is converging to a Gaussian in this case. The fact that $\tilde{g}(x)$ decays exponentially is also argued in [28] but the precise dependence of the constants on x_0, k_0, σ is not pinned down there (and the argument there does not use the Zak transform).

We state first a technical lemma.

Lemma 3.2.1 *Let* $\mathfrak{M} \in 2\mathbb{N}$. *Then* $S(x_0, \mathfrak{M}, \vec{t}, \vec{s})$ *reduces to:*

$$S(x_0, \mathfrak{M}, \vec{t}, \vec{s}) =$$

$$\left(\frac{\mathfrak{M}x_0}{\sqrt{\pi}} \right)^N \left(\sum_{\vec{l} \in \mathbb{Z}^N} \exp\left(-x_0^2[(\vec{s} - \vec{l})^2] \right) \right) \prod_{j=1}^{N} \theta_3(2\pi \mathfrak{M}\vec{t}_j | i x_0^2 \mathfrak{M}^2/4\pi) \qquad (3.2.1)$$

Proof Consider the sum in (3.1.6). We can compute:

$$\sum_{\vec{r}\in\{0,...\mathfrak{M}-1\}^N} \left| \sum_{\vec{l}\in\mathbb{Z}^N} \exp\left(2\pi i\vec{l}\cdot(\vec{t}-\vec{r}/\mathfrak{M})\right) \exp\left(\frac{-x_0^2}{2}(\vec{s}-\vec{l})^2\right) \right|^2 =$$

$$= \sum_{\vec{r}\in\{0,...\mathfrak{M}-1\}^N} \left[\sum_{\vec{l}\in\mathbb{Z}^N} \exp\left(2\pi i\vec{l}\cdot(\vec{t}-\vec{r}/\mathfrak{M})\right) \exp\left(\frac{-x_0^2}{2}(\vec{s}-\vec{l})^2\right) \right] \times$$

$$\left[\sum_{\vec{n}\in\mathbb{Z}^N} \exp\left(-2\pi i\vec{n}\cdot(\vec{t}-\vec{r}/\mathfrak{M})\right) \exp\left(\frac{-x_0^2}{2}(\vec{s}-\vec{n})^2\right) \right]$$

$$= \sum_{\vec{r}\in\{0,...\mathfrak{M}-1\}^N} \sum_{\vec{l}\in\mathbb{Z}^N} \sum_{\vec{n}\in\mathbb{Z}^N} \exp\left(2\pi i(\vec{l}-\vec{n})\cdot(\vec{t}-\vec{r}/\mathfrak{M})\right) \times$$

$$\exp\left(\frac{-x_0^2}{2}((\vec{s}-\vec{l})^2+(\vec{s}-\vec{n})^2)\right) \qquad (3.2.2)$$

For simplicity, in this calculation, $\vec{v}^2 = \sum_{j=1}^N \vec{v}_j^2$. Note that we do not take absolute values or complex conjugates anywhere, and thus our result is analytic. By passing the sum over $\vec{r}$ inside the other two sums, and noting the following:

$$\sum_{\vec{r}\in\{0,...\mathfrak{M}-1\}^N} \exp\left(-2\pi i(\vec{l}-\vec{n})\cdot(\vec{r}/\mathfrak{M})\right) = \begin{cases} 0, & (\vec{l}-\vec{n})\notin(\mathfrak{M}\mathbb{Z})^N \\ \mathfrak{M}^N, & (\vec{l}-\vec{n})\in(\mathfrak{M}\mathbb{Z})^N \end{cases}$$

We can then set $\vec{n}=\vec{l}+\mathfrak{M}\vec{k}$. We then find:

$$(3.2.2) = \sum_{\vec{l}\in\mathbb{Z}^N} \sum_{\vec{n}\in\mathbb{Z}^N} \sum_{\vec{r}\in\{0,...\mathfrak{M}-1\}^N}$$

$$\exp\left(2\pi i(\vec{l}-\vec{n})\cdot(\vec{t}-\vec{r}/\mathfrak{M})\right) \exp\left(\frac{-x_0^2}{2}((\vec{s}-\vec{l})^2+(\vec{s}-\vec{n})^2)\right) =$$

$$\mathfrak{M}^N \sum_{\vec{l}\in\mathbb{Z}^N} \sum_{\vec{k}\in\mathbb{Z}^N} \left[\exp\left(2\pi i(\mathfrak{M}\vec{k})\cdot(\vec{t}-\vec{r}/\mathfrak{M})\right) \right.$$

$$\left. \times \exp\left(\frac{-x_0^2}{2}((\vec{s}-\vec{l})^2+(\vec{s}-\vec{l}-\mathfrak{M}\vec{k})^2)\right) \right] =$$

$$\mathfrak{M}^N \sum_{\vec{k}\in\mathbb{Z}^N} \exp\left(2\pi i\mathfrak{M}\vec{k}\cdot\vec{t}\right) \sum_{\vec{l}\in\mathbb{Z}^N} \exp\left(-x_0^2[(\vec{s}-\vec{l}-\mathfrak{M}\vec{k}/2)^2+\mathfrak{M}^2\vec{k}^2/4]\right)$$

$$(3.2.3)$$

(3.2.3) holds whether $\mathfrak{M}$ is odd or even. For $\mathfrak{M}$ even, we find:

$$\sum_{\vec{l}\in\mathbb{Z}^N} \exp\left(-x_0^2[(\vec{s} - \vec{l} - \mathfrak{M}\vec{k}/2)^2 + \mathfrak{M}^2\vec{k}^2/4]\right) =$$

$$\exp(-x_0^2\mathfrak{M}^2\vec{k}^2/4) \sum_{\vec{l}\in\mathbb{Z}^N} \exp\left(-x_0^2[(\vec{s} - \vec{l})^2]\right)$$

This follows since the latter sum is merely an integer translate (in $\vec{l}$) of the former (since $\mathfrak{M}/2$ is an integer). But since the sum is taken over all $\mathbb{Z}^N$, integer translates do not matter. Then we can simplify (3.2.3) even further to:

(3.2.3)

$$= \mathfrak{M}^N \left(\sum_{\vec{l}\in\mathbb{Z}^N} \exp(-x_0^2[(\vec{s} - \vec{l})^2])\right) \sum_{\vec{k}\in\mathbb{Z}^N} \exp(2\pi i \mathfrak{M}\vec{k}\cdot\vec{t}) \exp(-x_0^2\mathfrak{M}^2\vec{k}^2/4)$$

$$= \mathfrak{M}^N \left(\sum_{\vec{l}\in\mathbb{Z}^N} \exp\left(-x_0^2(\vec{s} - \vec{l})^2\right)\right) \prod_{j=1}^{N} \theta_3(2\pi\mathfrak{M}\vec{t}_j | i x_0^2\mathfrak{M}^2\vec{k}^2/4\pi)$$

We now multiply by $(x_0/\sqrt{\pi})^N$ to recover $S(x_0, \mathfrak{M}, t, s)$, proving (3.2.1).   □

Theorem 3.2.2 *Let $x_0 k_0 = 2\pi/\mathfrak{M}$ for $\mathfrak{M} \in 2\mathbb{N}$. Let $\tilde{g}(\vec{x})$ be the dual window to the GWFT. Then $\tilde{g}(\vec{x})$ satisfies the following bounds:*

$$\|\tilde{g}(\vec{x})\|_{L^\infty} \leqslant \left(\frac{x_0}{\sigma}\right)^N \mathcal{A}_F^{-1} \sum_{\vec{n}\in\mathbb{Z}^N} \exp\left(-\frac{x_0^2|\vec{n}|^2}{\sigma^2}\right) = \left(\frac{x_0}{\sigma}\right)^N \mathcal{A}_F^{-1} \|\phi\|_{L^\infty(\vec{t},\vec{s})}$$

(3.2.4)

Letting $\vec{\alpha} = (\alpha_1, ...\alpha_N)$ be a multi-index, we find that:

$$\left|\partial_x^{\vec{\alpha}} \tilde{g}(\vec{x})\right| \leq \mathfrak{G}(x_0, k_0, N, \vec{\alpha}) e^{-\mathfrak{r}(x_0, k_0)|\vec{x}|_1}$$

(3.2.5)

*When s is a scalar, we will let $\mathfrak{G}(x_0, k_0, N, s) = \mathfrak{G}(x_0, k_0, N, (s, 0, .., 0))$.
The decay rate $\mathfrak{r}(x_0, k_0)$ is given by:*

$$\mathfrak{r}(x_0, k_0) = x_0\mathfrak{M}/8\pi\sigma$$

(3.2.6)

The constant $\mathfrak{G}(x_0, k_0, N, s)$ is defined below. We must first define the following auxiliary functions:

$$\phi(\vec{t}, \vec{s}) = [\mathbf{Z}e^{-x^2}](\vec{t}, \vec{s}) = \left(\frac{x_0}{\sigma}\right)^{N/2} \sum_{\vec{l} \in \mathbb{Z}^N} e^{2\pi i (\vec{t} \cdot \vec{l})} \exp(-x_0^2 |\vec{s} - \vec{l}|_{l^2}^2/\sigma^2) \quad (3.2.7)$$

$$F(x_0, \mathfrak{M}, t, x) = \frac{\phi(t - i\gamma, x/x_0)}{\mathfrak{M}x_0\pi^{-1/2} \left(\sum_{l \in \mathbb{Z}} e^{-(x-lx_0)^2/\sigma^2}\right) \theta_3(2\pi\mathfrak{M}t \,|\, ix_0^2\mathfrak{M}^2/4\pi\sigma^2)} \quad (3.2.8)$$

$$G(x_0, \mathfrak{M}, t, x) = \frac{\phi(t - i\gamma, x/x_0)}{\left(\sum_{l \in \mathbb{Z}} e^{-(x-lx_0)^2/\sigma^2}\right) \theta_{3,z}(2\pi\mathfrak{M}t_j \,|\, ix_0^2\mathfrak{M}^2/4\pi\sigma^2)} \quad (3.2.9)$$

Here, $\theta_3(z|\tau)$ is a Jacobi theta functions (see Definition 3.1.3). The notation $\theta_{3,z}$ signifies $\theta_{3,z}(z_0|\tau) = \partial_z\theta_3(z|\tau)|_{z=z_0}$.
We can now define the constant term:

$$\mathfrak{G}(x_0, k_0, N, \vec{\alpha}) = \prod_{j=1}^{N} \left(x_0^{1/2}\sigma^{-1/2}e^{x_0^2\mathfrak{M}/8\sigma^2} \left\|\partial_x^{\vec{\alpha}_j} F(x_0, \mathfrak{M}, t, x)\right\|_{L^\infty} \right.$$

$$\left. + \frac{\sigma^{1/2}}{2\mathfrak{M}^2 x_0^{1/2}\pi^{1/2}} \lfloor 2\pi\mathfrak{M} - 1/2 \rfloor \left\|\partial_x^{\vec{\alpha}_j} G(x_0, \mathfrak{M}, t, x)\right\|_{L^\infty} \right) \quad (3.2.10)$$

Proof In this theorem, we mainly do calculations on the dual window. We perform the calculations in 1 dimension, and then note that:

$$\tilde{g}(\vec{x}) = \prod_{j=1}^{N} \tilde{g}_{1D}(\vec{x}_j)$$

In one dimension, we find that (dropping the $1D$ subscript) the dual window can be computed (recalling that $\phi(t, s) = \mathbf{Z}e^{-x^2/2}$:

$$\tilde{g}(x) = \mathbf{Z}^{-1}\mathbf{Z}(F^*F)^{-1}e^{-x^2/2} = \mathbf{Z}^{-1}S(x_0, \mathfrak{M}, t, s)^{-1}\phi(t, s)$$

$$= x_0^{1/2} \int_0^1 \frac{e^{-i2\pi t \lfloor x/x_0 \rfloor}\phi(t, x/x_0)}{S(x_0, \mathfrak{M}, t, x/x_0)} dt \quad (3.2.11)$$

We also assume $\sigma = 1$, for simplicity. To do the calculation when $\sigma \neq 1$, we merely scale the result.

Bound in L^∞

To bound $\tilde{g}(x)$ in L^∞, we need only bound the integral. Note that $S(x_0, \mathfrak{M}, t, x/x_0)^{-1}$ is bounded by $\mathcal{A}_F^{-1}$ (by Theorem 3.1.4). Thus, we obtain the L^∞ bound:

$$\|\tilde{g}(x)\|_{L^\infty} \leq x_0^{1/2} \mathcal{A}_F^{-1} \|\phi(t,s)\|_{L^\infty([0,1]^2, dtds)}$$

Shifting the Integration Contour

Here we work in 1 space dimension. We then observe that the $\tilde{g}(\vec{x}) = \prod_{j=1}^{N} \tilde{g}_{1d}(\vec{x}_j)$.

To determine the decay of the dual window, we move the contour of integration in (3.2.11) up from $[0,1]$ to $[0,1] \pm i\gamma$ (depending on the sign of x, for simplicity we suppose $x > 0$). The constant is chosen to be $\gamma = i x_0^2 \mathfrak{M}/8\pi^2$, due to the fact that $\theta_3(z|\tau)$ obeys a recurrence relation with this period (see (3.1.5)).

The endpoints do not contribute to the integral, since $S(x_0, \mathfrak{M}, t, s)$ and $\phi(t, s)$ are 1-periodic in t. Thus, the integral in (3.2.11) becomes:

$$e^{-2\pi\gamma\lfloor x/x_0 \rfloor} x_0^{1/2} \int_0^1 \frac{e^{-i2\pi t \lfloor x/x_0 \rfloor} \phi(t - i\gamma, x/x_0)}{S(x_0, \mathfrak{M}, t - i\gamma, x/x_0)} dt + \text{Residues} \qquad (3.2.12)$$

Using (3.2.1) in one dimension, we find that:

$$S(x_0, \mathfrak{M}, t, s) = \mathfrak{M} x_0 \pi^{-1/2} \left(\sum_{l \in \mathbb{Z}} e^{-x_0^2(s-l)^2} \right) \theta_3(2\pi \mathfrak{M} t | i x_0^2 \mathfrak{M}^2/4\pi) \qquad (3.2.13)$$

We now need to find the zeros of $S(x_0, \mathfrak{M}, t, s)$ in the region $0 \leq \Re t \leq 1$, $0 \leq \Im t \leq \gamma$.

The product formula (3.1.4) for the function $\theta_3(z|\tau)$ implies that $\theta_3(z|\tau) = 0$ only when $(2n-1)i\pi\tau \pm 2\pi i z = -\pi i + 2\pi n i$ for some $n \in \mathbb{Z}$, and all zero's at these points are of first order.

Using this and (3.2.13), we find that the relevant zeros of $S(x_0, \mathfrak{M}, t, s)$ occur at $2\pi \mathfrak{M} t = 1/2 + j - i x_0^2 \mathfrak{M}^2/8\pi$, with $t \in [0,1]$. These are

$$t_j = (j + 1/2)/2\pi \mathfrak{M} + i x_0^2 \mathfrak{M}/16\pi^2$$

with $j = 0 \ldots \lfloor 2\pi \mathfrak{M} - 1/2 \rfloor$.

The residue term therefore takes the form:

$$\text{Residues} = x_0^{1/2} e^{-(x_0^2 \mathfrak{M}/16\pi)\lfloor x/x_0 \rfloor}$$

$$\times \sum_{j=0}^{\lfloor 2\pi \mathfrak{M} - 1/2 \rfloor} \frac{e^{-i2\pi t_j \lfloor x/x_0 \rfloor} \phi(t_j, x/x_0)}{\mathfrak{M} x_0 \pi^{-1/2} \left(\sum_{l \in \mathbb{Z}} e^{-(x - lx_0)^2} \right) 2\pi \mathfrak{M} \theta_{3,z}(2\pi \mathfrak{M} t_j | i x_0^2 \mathfrak{M}^2/4\pi)}$$

$$(3.2.14)$$

Here, $\theta_{3,z}(z_0|\tau) = \partial_z \theta_3(z|\tau)|_{z=z_0}$.

We combine these two results, and note that $\theta_3(z + \tau|\tau) = e^{-i\pi(\tau-2z)}\theta_3(z, \tau)$ to obtain the following expression for $\tilde{g}(x)$:

$$\tilde{g}(x) = e^{-(x_0^2\mathfrak{M}/8\pi)\lfloor x/x_0\rfloor}x_0^{1/2}e^{x_0^2\mathfrak{M}/8}\int_0^1 \frac{e^{-i2\pi t\lfloor x/x_0\rfloor}\phi(t-i\gamma, x/x_0)}{S(x_0, \mathfrak{M}, t, x/x_0)e^{i2\pi t}}dt$$

$$+ \frac{\pi^{-1/2}x_0^{-1/2}e^{-(x_0^2\mathfrak{M}/16\pi)\lfloor x/x_0\rfloor}}{2\mathfrak{M}^2\left(\sum_{l\in\mathbb{Z}}e^{-(x-lx_0)^2}\right)}\sum_{j=0}^{\lfloor 2\pi\mathfrak{M}-1/2\rfloor}\frac{e^{-i2\pi\mathfrak{M}t_j\lfloor x/x_0\rfloor}\phi(t_j, x/x_0)}{\theta_{3,z}(2\pi\mathfrak{M}t_j|ix_0^2\mathfrak{M}^2/4\pi)}$$

$$(3.2.15)$$

Calculation of Derivatives

Let us define the following two functions:

$$F(x_0, \mathfrak{M}, t, x) = \frac{\phi(t-i\gamma, x/x_0)}{S(x_0, \mathfrak{M}, t, x/x_0)}$$

$$= \frac{\phi(t-i\gamma, x/x_0)}{\mathfrak{M}x_0\pi^{-1/2}\left(\sum_{l\in\mathbb{Z}}e^{-(x-lx_0)^2}\right)\theta_3(2\pi\mathfrak{M}t|ix_0^2\mathfrak{M}^2/4\pi)}$$

$$G(x_0, \mathfrak{M}, t, x) = \frac{\phi(t-i\gamma, x/x_0)}{\left(\sum_{l\in\mathbb{Z}}e^{-(x-lx_0)^2}\right)\theta_{3,z}(2\pi\mathfrak{M}t_j|ix_0^2\mathfrak{M}^2/4\pi)}$$

Then we can rewrite (3.2.15) as follows:

$$\tilde{g}(x) = e^{-(x_0^2\mathfrak{M}/8\pi)\lfloor x/x_0\rfloor}x_0^{1/2}e^{x_0^2\mathfrak{M}/8}\int_0^1 e^{-i2\pi t\lfloor x/x_0\rfloor}F(x_0, \mathfrak{M}, t, x)dt$$

$$+ e^{-(x_0^2\mathfrak{M}/16\pi)\lfloor x/x_0\rfloor}\frac{\pi^{-1/2}x_0^{-1/2}}{2\mathfrak{M}^2}\sum_{j=0}^{\lfloor 2\pi\mathfrak{M}-1/2\rfloor}G(x_0, \mathfrak{M}, t, x) \qquad (3.2.16)$$

Calculation of the Decay Rate

Taking (3.2.16) as a starting point, we can now calculate the decay rate of $\tilde{g}(x)$. We use the simple fact that:

$$e^{-\alpha\lfloor x/x_0\rfloor} \le e^{\alpha}e^{-\alpha x/x_0} \qquad (3.2.17)$$

The decay rate can be computed simply enough, taking absolute values of (3.2.16) and using (3.2.17):

$$\left|\partial_x^n\tilde{g}(x)\right| \le e^{-(x_0\mathfrak{M}/8\pi)x}\left(x_0^{1/2}e^{x_0^2\mathfrak{M}/8}\left\|\partial_x^n F(x_0, \mathfrak{M}, t, x)\right\|_{L^\infty}\right.$$

$$\left. + \frac{\pi^{-1/2}x_0^{-1/2}}{2\mathfrak{M}^2}\lfloor 2\pi\mathfrak{M}-1/2\rfloor\left\|\partial_x^n G(x_0, \mathfrak{M}, t, x)\right\|_{L^\infty}\right)$$

This is what we wanted to prove. To obtain the result in N dimensions, we take products. To obtain the result when $\sigma \neq 1$, we scale.

$\square$

Corollary 3.2.3 *If we interchange $\vec{x}$ and $\vec{k}$, x_0 and k_0, and σ with σ^{-1} everywhere in the above theorem, then the conclusion still holds.*

Proof The Fourier transform of the WFT is still a WFT. The Fourier transform of the window function $e^{-|\vec{x}|^2/2}$ is $e^{-|\vec{k}|^2/2}$. Therefore the same result holds with $\vec{x}$ and $\vec{k}$ interchanged.

$\square$

3.3 Localization of Phases and Spaces

The WFT allows us to define a concrete realization of phase space. We will consider $\mathbb{Z}^N \times \mathbb{Z}^N$ to be a discrete realization of phase space, with the vector $(\vec{a}, \vec{b}) \in \mathbb{Z}^N \times \mathbb{Z}^N$ representing the point $\vec{a}x_0$ in position and $\vec{b}k_0$ in momentum. With this in mind, we can now construct phase space localization operators.

Definition 3.3.1 For a set $F \in \mathbb{Z}^N \times \mathbb{Z}^N$, we define the phase space localization operator:

$$\mathcal{P}_F \Psi(x) = \sum_{(\vec{a}, \vec{b}) \in F} \Psi_{(\vec{a}, \vec{b})} \phi_{(\vec{a}, \vec{b})}(\vec{x}) \qquad (3.3.1)$$

Intuitively, one expects that phase space localization based on the WFT will correspond to the usual phase space localization based on position and momentum projections. Of course, the correspondence is fuzzy, and we do make small errors (to be quantified shortly).

For convenience of notation, we make the following definition.

Definition 3.3.2 For $K \in \mathbb{R}^+$, we define the set of high frequency and low frequency framelets, respectively:

$$\mathrm{HF}(K) = \left\{ (\vec{a}, \vec{b}) \in \mathbb{Z}^N \times \mathbb{Z}^N : k_0 |\vec{b}|_\infty > K \right\} \qquad (3.3.2a)$$

$$\mathrm{LF}(K) = \left\{ (\vec{a}, \vec{b}) \in \mathbb{Z}^N \times \mathbb{Z}^N : k_0 |\vec{b}|_\infty \leq K \right\} \qquad (3.3.2b)$$

We now prove a result showing that a high pass filter constructed from the WFT is very similar to a high pass filter constructed from the Fourier transform.

Theorem 3.3.3 *Let $P^0_{B_{K'};k_0}(\vec{k})$ be a projection operator onto the set $[-(K - \mathfrak{K}^s(\epsilon)), K - \mathfrak{K}^s(\epsilon)]^N$. Then:*

$$\|\mathcal{P}_{\text{HF}(K)} f(x)\|_{H^s}$$

$$\leq \mathfrak{H}^s_+(\tilde{g}(\vec{x}))\mathfrak{H}^{-s}_+(e^{-x^2/\sigma^2})\left\|(1 - P^0_{B_{K'};k_0}(\vec{k}))\hat{f}(\vec{k})\right\|_{H^s} + \epsilon \|f(x)\|_{H^s} \qquad (3.3.3)$$

The constant $\mathfrak{K}^s(\epsilon)$ is defined by:

$$\mathfrak{K}^s(\epsilon) = \inf_{M \in \mathbb{N}}\left\{Mk_0 : \sqrt{\mathbf{h}_s^-}\,\mathfrak{G}(k_0, x_0, N, s)\right.$$

$$\left[(1 + \mathcal{L}_d)(x_0/2\pi)^{-N}(\mathbf{m}_{c,s}(\sigma, N) + \mathbf{m}'_{c,s}(\sigma, N)) + ((2 + \mathcal{L}_d)(x_0/2\pi)^{-N})\right]$$

$$\left. \times \mathcal{L}_s\left(1 + k_0^s \sum_{i=1}^{N}\left(\vec{z}_i \frac{d}{d\vec{z}_i}\right)^s\right) \mathbf{a}_{M,N}(\vec{z})\bigg|_{\vec{z}_j = e^{-\mathfrak{r}(k_0, x_0)k_0}} \leq \epsilon \right\}$$

$$= O(|\ln \epsilon|) \qquad (3.3.4)$$

with the generating function $\mathbf{a}_{M,s}(\vec{z})$ defined below, in Lemma 3.3.5.

Remark 3.3.4 We remark at this time that we do not believe our estimates are optimal. We have taken a number of shortcuts in the proofs of the various theorems in this section. We conjecture that these results can be improved significantly by a more careful analysis.

Before proceeding with the proof, we state a technical lemma.

Lemma 3.3.5 *We have the following bound for the discrete convolution:*

$$\sum_{\vec{a} \in \mathbb{Z}^N} \langle \vec{a}2\pi/x_0\rangle^s e^{-\sigma^2(\vec{a}2\pi/x_0 - \vec{z})^2}$$

$$\leq (1 + \mathcal{L}_d)(x_0/2\pi)^{-N}(\mathbf{m}_{c,s}(\sigma, N) + \mathbf{m}'_{c,s}(\sigma, N))$$

$$+ ((2 + \mathcal{L}_d)(x_0/2\pi)^{-N})\langle \vec{z}\rangle^s = O(\langle \vec{z}\rangle^s) \qquad (3.3.5a)$$

$$\sup_{|\vec{k}|_\infty < k_0} \sum_{|\vec{k} - \vec{b}k_0|_\infty \geq M} \langle \vec{b}k_0 - \vec{k}\rangle^s e^{-\mathfrak{r}(k_0, x_0)|\vec{b}k_0 - \vec{k}|_1}$$

$$\leq \mathcal{L}_s\left(1 + k_0^s \sum_{i=1}^{N}\left(\vec{z}_i \frac{d}{d\vec{z}_i}\right)^s\right) \mathbf{a}_{M,s}(\vec{z})\bigg|_{\vec{z}_j = e^{-\mathfrak{r}(k_0, x_0)k_0}}$$

$$= O(M^s e^{-\mathfrak{r}(k_0, x_0)M}) \qquad (3.3.5b)$$

The generating function $\mathbf{a}_{M,N}(\vec{z})$ *is defined as:*

$$\mathbf{a}_{M,N}(\vec{z}) = \left(\prod_{j=1}^{N} \frac{1}{1-\vec{z}_j}\right) \left[\sum_{1 \le j \le N} \vec{z}_j^M \left(1 + \sum_{j<i\le N} \vec{z}_i^M\right)\right] \tag{3.3.6}$$

Proof of Theorem 3.3.3 We proceed in three steps.

Setup

We begin by decomposing $f(x)$ into high and low frequencies, and applying the high pass filter:

$$\left\|\mathcal{P}_{HF(K)}f(x)\right\|_{H^s}$$
$$= \left\|\mathcal{P}_{HF(K)}P^0_{B_{K'};0}(\vec{k})f(x)\right\|_{H^s} + \left\|\mathcal{P}_{HF(K)}[1 - P^0_{B_{K'};0}(\vec{k})]f(x)\right\|_{H^s}$$

The first term is bounded by $\mathfrak{H}^s_+(\tilde{g}(\vec{x}))\mathfrak{H}^{-s}_+(e^{-x^2/\sigma^2})\left\|P^0_{B_{K'};0}(\vec{k})\hat{f}(\vec{k})\right\|_{H^s}$, thus it remains to bound the second. Let $h(x) \in H^{-s}$ have norm 1. Then:

$$\left\langle h(x)|\mathcal{P}_{HF(K)}[1 - P^0_{B_{K'};0}(\vec{k})]f(x)\right\rangle$$

$$= \sum_{(\vec{a},\vec{b})\in HF(K)} \left\langle h(x)|\phi_{(\vec{a},\vec{b})}(\vec{x})\right\rangle \left\langle \tilde{\phi}_{(\vec{a},\vec{b})}(\vec{x})|[1 - P^0_{B_{K'};0}(\vec{k})]f(x)\right\rangle$$

$$= \sum_{\vec{a}\in\mathbb{Z}^N} \sum_{k_0|\vec{b}|_\infty>K} \int\int \hat{h}(\vec{k})^* \hat{\phi}_{(\vec{a},\vec{b})}(\vec{k})\hat{\tilde{\phi}}_{(\vec{a},\vec{b})}(\vec{k}')^*[1 - P^0_{B_{K'};0}(\vec{k}')]\hat{f}(\vec{k}')d\vec{k}'d\vec{k}$$

$$= \int\int \left(\hat{h}(\vec{k})^*[1 - P^0_{B_{K'};0}(\vec{k}+\vec{z})]\hat{f}(\vec{k}+\vec{z})\right) \left(\sum_{\vec{a}\in\mathbb{Z}^N} e^{i\vec{a}x_0\cdot\vec{z}}\right)$$

$$\times \left(\sum_{k_0|\vec{b}|_\infty>K} \hat{\tilde{g}}(\vec{k}+\vec{z}-\vec{b}k_0)e^{-\sigma^2(\vec{k}-\vec{b}k_0)^2}\right) d\vec{z}d\vec{k} \tag{3.3.7}$$

Between lines 3 and 4 we changed variables to $\vec{k}' = \vec{k} + \vec{z}$, and used the identity:

$$\sum_{\vec{a}\in\mathbb{Z}^N} e^{i\vec{a}x_0\cdot\vec{z}} = \sum_{\vec{a}\in\mathbb{Z}^N} \delta(\vec{z} - \vec{a}2\pi/x_0)$$

Substituting this into (3.3.7) allows us to do the $\vec{z}$ integral. We obtain:

$$| (3.3.7)|^2 = \left| \int \hat{h}(\vec{k})^* \sum_{\vec{a} \in \mathbb{Z}^N} \sum_{k_0 |\vec{b}|_\infty > K} \left(\hat{\bar{g}}(\vec{k} + \vec{a} 2\pi/x_0 - \vec{b} k_0) e^{-\sigma^2 (\vec{k} - \vec{b} k_0)^2} \right. \right.$$

$$\left. \left. \times [1 - P^0_{B_{K'};0}(\vec{k} + \vec{a} 2\pi/x_0)] \hat{f}(\vec{k} + \vec{a} 2\pi/x_0) \right) d\vec{k} \right|^2$$

$$\leq \left\| \langle \vec{k} \rangle^{-s} \hat{h}(\vec{k}) \right\|^2_{L^2} \int \left| [1 - P^0_{B_{K'};0}(\vec{k})] \hat{f}(\vec{k}) \right.$$

$$\times \left. \sum_{\vec{a} \in \mathbb{Z}^N} \sum_{k_0 |\vec{b}|_\infty > K} \hat{\bar{g}}(\vec{k} - \vec{b} k_0) \langle \vec{k} - \vec{a} 2\pi/x_0 \rangle^s e^{-\sigma^2 (\vec{k} - \vec{a} 2\pi/x_0 - \vec{b} k_0)^2} \right|^2 d\vec{k}$$

$$\leq \mathbf{h}_s^- \| h(x) \|^2_{H^{-s}} \| f(x) \|^2_{L^2}$$

$$\times \sup_{|\vec{k}|_\infty \leq K - \mathfrak{K}^s(\epsilon)} \left| \sum_{\vec{a} \in \mathbb{Z}^N} \sum_{k_0 |\vec{b}|_\infty > K} \hat{\bar{g}}(\vec{k} - \vec{b} k_0) \langle \vec{k} - \vec{a} 2\pi/x_0 \rangle^s e^{-\sigma^2 (\vec{k} - \vec{a} 2\pi/x_0 - \vec{b} k_0)^2} \right|^2$$

$$(3.3.8)$$

Thus it remains to bound the sup term in the last equation.

Bounds on the Sum
We consider this term, dropping the $| \cdot |^2$ since everything underneath is positive.
We obtain:

$$\sup_{|\vec{k}|_\infty \leq K - \mathfrak{K}^s(\epsilon)} \sum_{\vec{a} \in \mathbb{Z}^N} \sum_{k_0 |\vec{b}|_\infty > K} \hat{\bar{g}}(\vec{k} - \vec{b} k_0) \langle \vec{k} - \vec{a} 2\pi/x_0 \rangle^s e^{-\sigma^2 (\vec{k} - \vec{a} 2\pi/x_0 - \vec{b} k_0)^2}$$

$$= \sup_{|\vec{k}|_\infty \leq K - \mathfrak{K}^s(\epsilon)} \sum_{k_0 |\vec{b}|_\infty > K} \hat{\bar{g}}(\vec{k} - \vec{b} k_0) \sum_{\vec{a} \in \mathbb{Z}^N} \langle \vec{k} - \vec{a} 2\pi/x_0 \rangle^s e^{-\sigma^2 (\vec{k} - \vec{a} 2\pi/x_0 - \vec{b} k_0)^2}$$

$$(3.3.9)$$

Thus we find, after applying Theorem 3.2.2 in order to bound the $\hat{\bar{g}}(\cdot)$ terms:

$$(3.3.9) \leq \sup_{|\vec{k}|_\infty \leq K - \mathfrak{K}^s(\epsilon)} \mathfrak{G}(k_0, x_0, N, s) \sum_{k_0 |\vec{b}|_\infty > K} e^{-\mathfrak{r}(k_0, x_0) |\vec{b} k_0 - \vec{k}|_1}$$

$$\times \sum_{\vec{a} \in \mathbb{Z}^N} \langle \vec{k} - \vec{a} 2\pi/x_0 \rangle^s e^{-\sigma^2 (\vec{k} - \vec{a} 2\pi/x_0 - \vec{b} k_0)^2} \qquad (3.3.10)$$

We bound the sum over $\vec{a}$ term using Lemma 3.3.5 (stated just after this proof), in particular (3.3.5a). This yields:

$$(3.3.10) \leq \mathfrak{G}(k_0, x_0, N, s)(1 + \mathfrak{L}_d)(x_0/2\pi)^{-N}(\mathbf{m}_{c,s}(\sigma, N) + \mathbf{m}'_{c,s}(\sigma, N))$$

$$\times \sup_{\left|\vec{k}\right|_\infty < K - \mathfrak{K}^s(\epsilon)} \sum_{k_0\left|\vec{b}\right|_\infty > K} e^{-\mathfrak{r}(k_0, x_0)\left|\vec{b}k_0 - \vec{k}\right|_1}$$

$$+ \mathfrak{G}(k_0, x_0, N, s)((2 + \mathfrak{L}_d)(x_0/2\pi)^{-N})$$

$$\times \sup_{\left|\vec{k}\right|_\infty < K - \mathfrak{K}^s(\epsilon)} \sum_{k_0\left|\vec{b}\right|_\infty > K} \langle\vec{b}k_0 - \vec{k}\rangle^s e^{-\mathfrak{r}(k_0, x_0)\left|\vec{b}k_0 - \vec{k}\right|_1} \qquad (3.3.11)$$

We observe now that for $|\vec{k}|_\infty \leq K - \mathfrak{K}^s(\epsilon)$, we find that $|\vec{k} - \vec{b}k_0|_\infty \geq \mathfrak{K}^s(\epsilon)$ if $k_0|\vec{b}|_\infty \geq K$. Therefore:

$$(3.3.11) \leq \mathfrak{G}(k_0, x_0, N, s)(1 + \mathfrak{L}_d)(x_0/2\pi)^{-N}(\mathbf{m}_{c,s}(\sigma, N) + \mathbf{m}'_{c,s}(\sigma, N))$$

$$\times \sup_{\left|\vec{k}\right|_\infty < K - \mathfrak{K}^s(\epsilon)} \sum_{|\vec{k} - \vec{b}k_0|_\infty \geq \mathfrak{K}^s(\epsilon)} e^{-\mathfrak{r}(k_0, x_0)\left|\vec{b}k_0 - \vec{k}\right|_1}$$

$$+ \mathfrak{G}(k_0, x_0, N, s)((2 + \mathfrak{L}_d)(x_0/2\pi)^{-N})$$

$$\times \sup_{\left|\vec{k}\right|_\infty < K - \mathfrak{K}^s(\epsilon)} \sum_{|\vec{k} - \vec{b}k_0|_\infty \geq \mathfrak{K}^s(\epsilon)} \langle\vec{b}k_0 - \vec{k}\rangle^s e^{-\mathfrak{r}(k_0, x_0)\left|\vec{b}k_0 - \vec{k}\right|_1}$$

$$\leq \mathfrak{G}(k_0, x_0, N, s)\Big[(1 + \mathfrak{L}_d)(x_0/2\pi)^{-N}(\mathbf{m}_{c,s}(\sigma, N) + \mathbf{m}'_{c,s}(\sigma, N))$$

$$+ ((2 + \mathfrak{L}_d)(x_0/2\pi)^{-N})\Big]$$

$$\times \sup_{\left|\vec{k}\right|_\infty < k_0} \sum_{|\vec{k} - \vec{b}k_0|_\infty \geq \mathfrak{K}^s(\epsilon)} \langle\vec{b}k_0 - \vec{k}\rangle^s e^{-\mathfrak{r}(k_0, x_0)\left|\vec{b}k_0 - \vec{k}\right|_1} \qquad (3.3.12)$$

To get from the first inequality to the second, we used the fact that $\langle\vec{b}k_0 - \vec{k}\rangle^s \geq 1$ to combine the sums.[1] Then we used the fact that the sum is invariant under translations on the lattice $k_0\mathbb{Z}^N$ to reduce the domain of the sup.

---

[1] This is a suboptimal result, but differs from the best result based on this proof strategy only logarithmically.

We bound this (applying 3.3.5, in particular (3.3.5b)) as follows:

$$(3.3.12) \le \mathfrak{G}(k_0, x_0, N, s)\Big[(1 + \mathcal{L}_d)(x_0/2\pi)^{-N}(\mathbf{m}_{c,s}(\sigma, N) + \mathbf{m}'_{c,s}(\sigma, N))$$

$$+ ((2 + \mathcal{L}_d)(x_0/2\pi)^{-N})\Big]$$

$$\times \mathcal{L}_s \left(1 + k_0^s \sum_{i=1}^{N} \left(\bar{z}_i \frac{d}{d\bar{z}_i}\right)\right)^s \mathbf{a}_{\mathfrak{R}^s(\epsilon),s}(\bar{z})\Big|_{\bar{z}_j = e^{-\mathfrak{r}(k_0,x_0)k_0}} \tag{3.3.13}$$

We note that the bound in (3.3.13) is $O(\mathfrak{R}^s(\epsilon)^2 e^{-\mathfrak{r}(k_0,x_0)\mathfrak{R}^s(\epsilon)})$.

Conclusion

We now finish the argument. We observe that (by (3.3.7) and (3.3.8)):

$$\Big|\big\langle h(x)|\mathcal{P}_{\mathrm{HF}(K)}[1 - P^0_{B_{K'};0}(\vec{k})]f(x)\big\rangle\Big|^2$$

$$\le \mathbf{h}_s^- \,\|h(x)\|^2_{H^{-s}} \,\|f(x)\|^2_{L^2} \,|(3.3.13)|^2 \tag{3.3.14}$$

for any $h(x)$ having norm 1 in H^{-s}. Thus:

$$\Big\|\mathcal{P}_{\mathrm{HF}(K)}[1 - P^0_{B_{K'};0}(\vec{k})]f(x)\Big\|_{H^s} \le \sqrt{\mathbf{h}_s^-}\,\|f(x)\|_{H^s}\,|(3.3.13)|$$

$$= \left(\sqrt{\mathbf{h}_s^-}\,\mathfrak{G}(k_0, x_0, N, s)\Big[(1 + \mathcal{L}_d)(x_0/2\pi)^{-N}(\mathbf{m}_{c,s}(\sigma, N) + \mathbf{m}'_{c,s}(\sigma, N))\right.$$

$$+ ((2 + \mathcal{L}_d)(x_0/2\pi)^{-N})\Big]$$

$$\left.\times \mathcal{L}_s \left(1 + k_0^s \sum_{i=1}^{N} \left(\bar{z}_i \frac{d}{d\bar{z}_i}\right)\right)^s \mathbf{a}_{\mathfrak{R}^s(\epsilon),s}(\bar{z})\Big|_{\bar{z}_j = e^{-\mathfrak{r}(k_0,x_0)k_0}}\right) \|f(x)\|_{H^s}$$

$$= O(\mathfrak{R}^s(\epsilon)^2 e^{-\mathfrak{r}(k_0,x_0)\mathfrak{R}^s(\epsilon)})\,\|f(x)\|_{H^s}$$

But $\mathfrak{R}^s(\epsilon)$ is defined precisely so that this is less than $\epsilon\,\|f(x)\|_{H^s}$.

$\square$

Proof of Lemma 3.3.5 Divide and conquer.

Equation (3.3.5a) We interpret this as a Riemann sum approximating an integral, and calculate.

$$\sum_{\vec{a}\in\mathbb{Z}^N} \langle \vec{a}2\pi/x_0\rangle^s e^{-\sigma^2(\vec{a}2\pi/x_0 - \bar{z})^2}$$

$$\le (x_0/2\pi)^{-N}\left(\int_{\mathbb{R}^N} \langle \vec{a}\rangle^s e^{-\sigma^2(\vec{a}-\bar{z})^2} + \Big|\nabla\langle \vec{a}\rangle^s e^{-\sigma^2(\vec{a}-\bar{z})^2}\Big|_1 d\vec{a}\right)$$

$$\leq (x_0/2\pi)^{-N} \int_{\mathbb{R}^N} \langle \vec{a} \rangle^s e^{-\sigma^2 (\vec{a} - \vec{z})^2} d\vec{a} + (x_0/2\pi)^{-N} \int_{\mathbb{R}^N} \left| \nabla \langle \vec{a} \rangle^s \right| e^{-\sigma^2 (\vec{a} - \vec{z})^2} d\vec{a}$$

$$+ \int_{\mathbb{R}^N} \langle \vec{a} \rangle^s \left| \nabla e^{-\sigma^2 (\vec{a} - \vec{z})^2} \right| d\vec{a}$$

$$\leq (1 + \mathcal{L}_d)(x_0/2\pi)^{-N} \int_{\mathbb{R}^N} \langle \vec{a} \rangle^s e^{-\sigma^2 (\vec{a} - \vec{z})^2} d\vec{a}$$

$$+ (x_0/2\pi)^{-N} \int_{\mathbb{R}^N} \langle \vec{a} \rangle^s \left| \nabla e^{-\sigma^2 (\vec{a} - \vec{z})^2} \right| d\vec{a}$$

$$\leq (1 + \mathcal{L}_d)(x_0/2\pi)^{-N} (\mathbf{m}_{c,s}(\sigma, N) + \mathbf{m}'_{c,s}(\sigma, N)) + ((2 + \mathcal{L}_d)(x_0/2\pi)^{-N}) \langle \vec{z} \rangle^s$$

Equation (3.3.5b) First we consider the sum over $\vec{b}_j \geq 0$ only, and pull out a factor of 2^N:

$$\sum_{|\vec{k} - \vec{b}k_0|_\infty \geq M} \langle \vec{b}k_0 - \vec{k} \rangle^s e^{-\tau(k_0, x_0) |\vec{b}k_0 - \vec{k}|_1}$$

$$\leq 2^N \sum_{\substack{|\vec{k} - \vec{b}k_0|_\infty \geq M \\ \vec{b}_j \geq 0}} \langle \vec{b}k_0 - \vec{k} \rangle^s e^{-\tau(k_0, x_0) |\vec{b}k_0 - \vec{k}|_1}$$

$$\leq 2^N \sum_{\substack{|\vec{b}k_0|_\infty \geq M - k_0 \\ \vec{b}_j \geq 0}} \mathcal{L}_s (1 + |\vec{b}k_0|_s^s) e^{-\tau(k_0, x_0) |\vec{b}k_0|_1} \qquad (3.3.15)$$

The last line follows because $|\vec{k}|_\infty \leq k_0$. We will now represent the sum by a generating function, analytic jointly in the variable $\vec{z}$. We will evaluate the generating function at $\vec{z}_j = e^{-\tau(k_0, x_0)k_0}$ to obtain the bound.

Note that:

$$\sum_{\substack{|\vec{b}|_\infty > M \\ \vec{b}_j \geq 0}} \vec{z}^{\vec{b}} = \sum_{\vec{b}_j \geq 0} \vec{z}^{\vec{b}} - \sum_{\substack{|\vec{b}|_\infty \leq M \\ \vec{b}_j > 0}} \vec{z}^{\vec{b}} = \prod_{j=1}^{N} \frac{1}{1 - \vec{z}_j} - \prod_{j=1}^{N} \left(\frac{1 - \vec{z}_j^M}{1 - \vec{z}_j} \right)$$

$$= \left(\prod_{j=1}^{N} \frac{1}{1 - \vec{z}_j} \right) \left[\sum_{1 \leq j \leq N} \vec{z}_j^M \left(1 + \sum_{j < i \leq N} \vec{z}_i^M \right) \right] = \mathbf{a}_{M,s}(\vec{z})$$

Then observe that multiplying under the sum by $\vec{k}_i^s$ is equivalent to applying the operator $\vec{z}_i \frac{d}{dz_i}$ to the generating function. Thus:

$$\sum_{\substack{|\vec{b}|_\infty > K \\ \vec{b}_j \geq 0}} \langle \vec{b}k_0 \rangle^s e^{-\mathfrak{r}(k_0,x_0)|\vec{b}k_0|_1} \leq \sum_{\substack{|\vec{b}|_\infty > K \\ \vec{b}_j \geq 0}} \mathfrak{L}_s \left(1 + k_0^s \sum_{i=1}^N \vec{b}_i^s \right) e^{-\mathfrak{r}(k_0,x_0)|\vec{b}k_0|_1}$$

$$= \mathfrak{L}_s \left(1 + k_0^s \sum_{i=1}^N \left(\vec{z}_i \frac{d}{dz_i} \right)^s \right) \mathbf{a}_{M,s}(\vec{z}) \Big|_{\vec{z}_j = e^{-\mathfrak{r}(k_0,x_0)k_0}}$$

$$= O(M^s e^{-\mathfrak{r}(k_0,x_0)M})$$

Thus we obtain the bound we seek.

$\square$

Remark 3.3.6 When we get around to stating our assumptions, one of them (Assumption 2) will demand that $\left\| \mathcal{P}_{HF(K)} f(x) \right\|_{H^s}$ be small. That formulation is merely for technical simplicity, and Theorem 3.3.3 will allow us to use the simpler

statement that $\left\| P^0_{|\vec{k}|_\infty > K; k_0}(\vec{k}) f(x) \right\|_{H^s}$ is small instead.

We now state a theorem regarding the phase space localization of the Gaussian WFT. The theorem says that if function $f(x)$ is small outside the box $[-X, X]^N \times [-K, K]^N$ (in phase space), then $f_{(\vec{a},\vec{b})}$ are small outside a somewhat larger box

$$[-X - \mathfrak{X}^s(\epsilon, K), X + \mathfrak{X}^s(\epsilon, K)] \times [-K - \mathfrak{K}^s(\epsilon, K), K + \mathfrak{K}^s(\epsilon, K)]$$

(with $\mathfrak{X}^s(\epsilon, K)$ and $\mathfrak{K}^s(\epsilon, K)$ given below).

This result is similar to Theorem 3.5.2 from [28]. Our result applies to N dimensions and an arbitrary Sobolev space (rather than 1 dimension and L^2) and compute the constants explicitly, however the result in [28] applies to an arbitrary window and oversampling rate.

Theorem 3.3.7 *Let* $B_X = [-X, X]^N$, $B_K = [-K, K]^N$ *for* $X, K < \infty$. *Then letting* $X' = X - \mathfrak{X}_\square^s(\epsilon, K, X)$, $K' = K - \mathfrak{K}_\square^s(\epsilon, K)$, *we find that:*

$$\left\| f(x) - \mathcal{P}_{B_{X'} \times B_{K'}} f(x) \right\|_{H^s} \leq \mathfrak{H}_+^s(\tilde{g}(\vec{x})) \mathfrak{H}_+^{-s}(e^{-x^2/\sigma^2})$$

$$\times \left(\left\| (1 - P^s_{B_X; x_0}(\vec{x})) f(\vec{x}) \right\|_{H^s} + \left\| (1 - P^0_{B_K; k_0}(\vec{k})) f(\vec{x}) \right\|_{H^s} + \epsilon \|f\|_{H^s} \right)$$

$$\tag{3.3.16}$$

The constants are given by:

$$\mathfrak{X}_{\square}^s(\epsilon, K, X) = \inf \Big\{ t \in \mathbb{R}^+ :$$

$$e^{-\mathfrak{r}(x_0,k_0)t} \sum_{j=0}^{\infty} 2N(2j + 2\lceil(X + t/x_0\rceil + 1)^{N-1} e^{-\mathfrak{r}(x_0,k_0)j}$$

$$\leq (\epsilon/2) \times \Big[\mathfrak{G}(x_0, k_0, N, 0) 2^{(N+2)/2}(X + x_0)^{(N-1)/2}$$

$$\times \sum_{|\vec{b}|_\infty \leq K'/k_0} \left(1 + \mathfrak{F}_s^2 \left(\sum_{k=1}^{N}(2\sigma)^{-s}(\exp_s(\sqrt{2\sigma}\vec{b}_k k_0))^2\right)\right)^{1/2}\Big]^{-1}\Big\} \qquad (3.3.17\text{a})$$

$$\mathfrak{K}_{\square}^s(\epsilon, K) = \mathfrak{K}^s(\epsilon/2) \qquad (3.3.17\text{b})$$

Proof **Setup**
To begin, we break this up into two separate problems:

$$\left\| f(x) - \mathcal{P}_{B_{X'} \times B_{K'}} \right\|_{H_b^s} = \left\| (\mathcal{P}_{\text{HF}(K')} + \mathcal{P}_{\text{LF}(K') \cap B_{X'}^C}) f(x) \right\|_{H_b^s}$$

$$\leq \left\| \mathcal{P}_{\text{HF}(K')} f(x) \right\|_{H_b^s} + \left\| \mathcal{P}_{\text{LF}(K') \cap B_{X'}^C} f(x) \right\|_{H_b^s}$$

$$\leq \left\| \mathcal{P}_{\text{HF}(K')} f(x) \right\|_{H_b^s}$$

$$+ \left\| \mathcal{P}_{\text{LF}(K') \cap B_{X'}^C} (1 - P_{B_X;x_0}^s(\vec{x})) f(x) \right\|_{H_b^s} \left\| \mathcal{P}_{\text{LF}(K') \cap B_{X'}^C} P_{B_X;x_0}^s(\vec{x}) f(x) \right\|_{H_b^s}$$

We apply Theorem 3.3.3 to $\left\| \mathcal{P}_{\text{HF}(K')} f(x) \right\|_{H_b^s}$, and bound

$$\left\| \mathcal{P}_{\text{LF}(K') \cap B_{X'}^C} (1 - P_{B_X;x_0}^s(\vec{x})) f(x) \right\|_{H_b^s}$$

$$\leq \mathfrak{H}_+^s(\tilde{g}(\vec{x})) \mathfrak{H}_+^{-s}(e^{-x^2/\sigma^2}) \left\| (1 - P_{B_X;x_0}^s(\vec{x})) f(x) \right\|_{H_b^s},$$

obtaining:

$$\left\| f(x) - \mathcal{P}_{B_{X'} \times B_{K'}} \right\|_{H_b^s}$$

$$\leq \mathfrak{H}_+^s(\tilde{g}(\vec{x})) \mathfrak{H}_+^{-s}(e^{-x^2/\sigma^2}) \left\| (1 - P_{B_K;k_0}^0(\vec{k})) \hat{f}(\vec{k}) \right\|_{H^s} + (\epsilon/2) \left\| f(x) \right\|_{H^s}$$

$$+ \mathfrak{H}^s_+(\tilde{g}(\vec{x})) \mathfrak{H}^{-s}_+ (e^{-x^2/\sigma^2}) \left\| (1 - P^s_{BX;x_0}(\vec{x})) f(x) \right\|_{H^s_b}$$

$$+ \left\| \mathcal{P}_{\mathrm{LF}(K') \cap B^C_{X'}} f(x) \right\|_{H^s_b}$$

Thus, to complete the proof, we must bound the last term by $(\epsilon/2) \| f(x) \|_{H^s_b}$. We write:

$$\mathcal{P}_{\mathrm{LF}(K') \cap B^C_{X'}} P^s_{BX;x_0}(\vec{x}) f(x)$$

$$= \sum_{|\vec{a}|_\infty > X'/x_0} \sum_{|\vec{b}|_\infty \leq K'/k_0} \left\langle \tilde{\phi}_{(\vec{a},\vec{b})}(\vec{x}) | P^s_{BX;x_0}(\vec{x}) f(x) \right\rangle \phi_{(\vec{a},\vec{b})}(\vec{x})$$

$$= \sum_{|\vec{a}|_\infty > X'/x_0} \sum_{|\vec{b}|_\infty \leq K'/k_0} \phi_{(\vec{a},\vec{b})}(\vec{x}) \int_{\mathbb{R}^N} e^{-i\vec{b}k_0 \cdot \vec{x}} \tilde{g}(\vec{x} - \vec{a}x_0)^* P^s_{BX;x_0}(\vec{x}) f(\vec{x}) d\vec{x}$$

$$(3.3.18)$$

We will construct first a bound on the integral term, as a function of $\vec{a}, \vec{b}$, and then return to (3.3.18) to complete the proof.

Bounds per Framelet
For $\vec{b}$ small, we do the following:

$$\left| \int_{\mathbb{R}^N} e^{-i\vec{b}k_0 \cdot \vec{x}} \tilde{g}(\vec{x} - \vec{a}x_0)^* P^s_{BX;x_0}(\vec{x}) f(\vec{x}) d\vec{x} \right|$$

$$\leq \int_{\mathbb{R}^N} \mathfrak{G}(x_0, k_0, N, 0) \left| e^{-\mathfrak{r}(x_0,k_0)|\vec{x} - \vec{a}x_0|_1} P^s_{BX;x_0}(\vec{x}) f(\vec{x}) \right| d\vec{x} \qquad (3.3.19)$$

Observe that $D = [-(X + x_0), (X + x_0)]^N$ contains the support of $P^s_{BX;x_0}(\vec{x})$, and apply Cauchy-Schwartz to obtain:

$$| (3.3.19) | \leq \mathfrak{G}(x_0, k_0, N, 0) \left\| P^0_{D;0}(\vec{x}) e^{-\mathfrak{r}(x_0,k_0)|\vec{x} - \vec{a}x_0|_1} \right\|_{L^2} \| f(\vec{x}) \|_{L^2} \qquad (3.3.20)$$

We now wish to bound $\left\| P^0_{D;0}(\vec{x}) e^{-\mathfrak{r}(x_0,k_0)|\vec{x} - \vec{a}x_0|_1} \right\|_{L^2}$.

We assume, without loss of generality, that $\vec{a}_j \geq 0$ for $j = 1..N$. We also observe that $|\vec{a}x_0|_\infty \geq (X + x_0)$, and let l be the (possibly not unique) dimension in $|\vec{a}_l x_0| - |\vec{x}_l|$ is maximized. Then:

$$\left\| P^0_{D;0}(\vec{x}) e^{-\mathfrak{r}(x_0,k_0)|\vec{x} - \vec{a}x_0|_1} \right\|_{L^2} = \left(\int_{[-(X+x_0), X+x_0]^N} e^{-2\mathfrak{r}(x_0,k_0)|\vec{x} - \vec{a}x_0|_1} d\vec{x} \right)^{1/2}$$

$$\leq \left(2^N \int_{[0, X+x_0]^N} e^{-2\mathfrak{r}(x_0,k_0)|\vec{x} - \vec{a}x_0|_1} d\vec{x} \right)^{1/2}$$

$$\leq \left(2^N \int_{[0,X+x_0]^N} e^{-2\mathfrak{r}(x_0,k_0)|\vec{x}-\vec{a}x_0|_\infty} d\vec{x} \right)^{1/2}$$

$$\leq 2^{N/2} \left(\int_{[0,X+x_0]} \int_{[0,X+x_0]^{N-1}} e^{-2\mathfrak{r}(x_0,k_0)(\vec{a}_l-\vec{x}_l)} d\vec{x}_\perp d\vec{x}_l \right)^{1/2}$$

$$= 2^{N/2}(X+x_0)^{(N-1)/2} \left(e^{-\mathfrak{r}(x_0,k_0)(\vec{a}_l-(X+x_0))} - e^{-\mathfrak{r}(x_0,k_0)\vec{a}_l} \right)$$

$$\leq 2^{(N+2)/2}(X+x_0)^{(N-1)/2} e^{-\mathfrak{r}(x_0,k_0)|\vec{a}|_\infty} (1+e^{\mathfrak{r}(x_0,k_0)(X+x_0)}) \qquad (3.3.21)$$

Noting that $(1+e^{\mathfrak{r}(x_0,k_0)(X+x_0)}) \leq 2e^{\mathfrak{r}(x_0,k_0)(x+x_0)}$, we find:

$$\left| \int_{\mathbb{R}^N} e^{-i\vec{b}k_0\cdot\vec{x}} \tilde{g}(\vec{x}-\vec{a}x_0)^* P^s_{BX:x_0}(\vec{x}) f(\vec{x}) d\vec{x} \right|$$

$$\leq (\mathfrak{G}(x_0,k_0,N,0)2^{(N+4)/2})$$

$$\times \|f(\vec{x})\|_{L^2} (X+x_0)^{(N-1)/2} e^{-\mathfrak{r}(x_0,k_0)|\vec{a}|_\infty} e^{\mathfrak{r}(x_0,k_0)(X+x_0)} \qquad (3.3.22)$$

Conclusion
We now return to (3.3.18).

$$\| (3.3.18)\|_{H^s} \leq \sum_{|\vec{a}|_\infty > X'/x_0} \sum_{|\vec{b}|_\infty \leq K'/k_0}$$

$$\left\| \phi_{(\vec{a},\vec{b})}(\vec{x}) \right\|_{H^s} \left| \int_{\mathbb{R}^N} e^{-i\vec{b}k_0\cdot\vec{x}} \tilde{g}(\vec{x}-\vec{a}x_0)^* P^s_{BX:x_0}(\vec{x}) f(\vec{x}) d\vec{x} \right|$$

$$\leq \sum_{|\vec{a}|_\infty > X'/x_0} \sum_{|\vec{b}|_\infty \leq K'/k_0} \left[\left(1 + \mathfrak{F}^2_s \left(\sum_{k=1}^N (2\sigma)^{-s} (\exp_s(\sqrt{2\sigma}\vec{b}_k k_0))^2 \right) \right) \right]^{1/2}$$

$$\times \|f(\vec{x})\|_{L^2} \mathfrak{G}(x_0,k_0,N,0)2^{(N+2)/2}$$

$$\times (X+x_0)^{(N-1)/2} e^{-\mathfrak{r}(x_0,k_0)|\vec{a}|_\infty} e^{\mathfrak{r}(x_0,k_0)(X+x_0)} \Bigg] \qquad (3.3.23)$$

To get from the second line to the third line, we applied Proposition 3.1.9 to bound $\left\| \phi_{(\vec{a},\vec{b})}(\vec{x}) \right\|_{H^s}$ and (3.3.22) to bound the integral term in the second line.

We now do the sum over $\vec{b}$ first, pulling out the terms that depend only on $\vec{b}$:

$$(3.3.23) \leq \left(\sum_{|\vec{b}|_\infty \leq K'/k_0} \left(1 + \mathfrak{F}_s^2 \left(\sum_{k=1}^N (2\sigma)^{-s} (\exp_s(\sqrt{2\sigma}\vec{b}_k k_0))^2 \right) \right)^{1/2} \right)$$

$$\times \|f(\vec{x})\|_{L^2} \, \mathfrak{G}(x_0, k_0, N, 0) 2^{(N+2)/2}(X + x_0)^{(N-1)/2}$$

$$\times \left(e^{\tau(x_0,k_0)(X+x_0)} \sum_{|\vec{a}|_\infty > X'/x_0} e^{-\tau(x_0,k_0)|\vec{a}|_\infty} \right) \tag{3.3.24}$$

We observe that for a given integer j, the number of integer lattice pts $\vec{a}$ with $|\vec{a}|_\infty = j$ is bounded by $2N(2j+1)^{N-1}$. We also note that $X' = X + \mathfrak{X}_\square^s(\epsilon, K, X)$, to find:

$$e^{\tau(x_0,k_0)(X+x_0)} \sum_{|\vec{a}|_\infty > X'/x_0} e^{-\tau(x_0,k_0)|\vec{a}|_\infty}$$

$$= e^{\tau(x_0,k_0)(X+x_0)} \sum_{j > (X+\mathfrak{X}_\square^s(\epsilon,K,X))/x_0} 2N(2j+1)^{N-1} e^{-\tau(x_0,k_0)j}$$

$$= e^{\tau(x_0,k_0)(X+x_0)} e^{-\tau(x_0,k_0)\lceil (X+\mathfrak{X}_\square^s(\epsilon,K,X))/x_0 \rceil x_0}$$

$$\times \sum_{j=0}^\infty 2N(2j + 2\lceil (X + \mathfrak{X}_\square^s(\epsilon, K, X))/x_0 \rceil + 1)^{N-1} e^{-\tau(x_0,k_0)j}$$

$$\leq e^{-\tau(x_0,k_0)\mathfrak{X}_\square^s(\epsilon,K,X)} \sum_{j=0}^\infty 2N(2j+2\lceil (X+\mathfrak{X}_\square^s(\epsilon, K, X))/x_0 \rceil+1)^{N-1} e^{-\tau(x_0,k_0)j}$$

By the definition of X', we find that:

$$e^{-\tau(x_0,k_0)\mathfrak{X}_\square^s(\epsilon,K,X)}$$

$$\times \sum_{j=0}^\infty 2N(2j + 2\lceil (X + \mathfrak{X}_\square^s(\epsilon, K, X))/x_0 \rceil + 1)^{N-1} e^{-\tau(x_0,k_0)j}$$

$$\leq (\epsilon/2) \times \left[\mathfrak{G}(x_0, k_0, N, 0) 2^{(N+2)/2}(X + x_0)^{(N-1)/2} \right.$$

$$\left. \times \sum_{|\vec{b}|_\infty \leq K'/k_0} \left(1 + \mathfrak{F}_s^2 \left(\sum_{k=1}^N (2\sigma)^{-s}(\exp_s(\sqrt{2\sigma}\vec{b}_k k_0))^2 \right) \right)^{1/2} \right]^{-1}$$

and therefore

$$(3.3.24) \leq (\epsilon/2) \, \|f(x)\|_{L^2} \tag{3.3.25}$$

Thus, we observe that:

$$\left\| \mathcal{P}_{\mathrm{LF}(K') \cap B_{X'}^C} P^s_{B_X; x_0}(\vec{x}) f(x) \right\|_{H^s} = \| (3.3.18) \|_{H^s} \leq (3.3.23) \leq (3.3.24) \leq (3.3.25)$$

$$\leq (\epsilon/2) \, \|f(x)\|_{L^2} \leq (\epsilon/2) \, \|f(x)\|_{H^s}$$

This is what we wanted to prove (recalling the discussion just before (3.3.18)).

$\square$

Remark 3.3.8 One can tune this estimate more carefully, making $\mathfrak{X}^s_\square(\epsilon, K, X)$ smaller at the cost of a larger $\mathfrak{K}^s_\square(\epsilon, K)$. For $\theta \in (0, 1)$, the following choices of $\mathfrak{X}^s_\square(\epsilon, K, X)$ and $\mathfrak{K}^s_\square(\epsilon, K)X$ are also valid:

$$\mathfrak{X}^s_\square(\epsilon, K, X) = \inf \left\{ t \in \mathbb{R}^+ : \right.$$

$$e^{-\mathfrak{r}(x_0, k_0)t} \sum_{j=0}^{\infty} 2N(2j + 2\lceil (X + t/x_0] + 1)^{N-1} e^{-\mathfrak{r}(x_0, k_0)j}$$

$$\leq \epsilon \theta \left[\mathfrak{G}(x_0, k_0, N, 0) 2^{(N+2)/2} (X + x_0)^{(N-1)/2} \right.$$

$$\left. \times \sum_{|\vec{b}|_\infty \leq K'/k_0} \left(1 + \mathfrak{F}^2_s \left(\sum_{k=1}^{N} (2\sigma)^{-s} (\exp_s(\sqrt{2\sigma} \vec{b}_k k_0))^2 \right) \right)^{1/2} \right]^{-1} \right\}$$

$$\mathfrak{K}^s_\square(\epsilon, K) = \mathfrak{K}^s(\epsilon(1 - \theta)) \tag{3.3.26a}$$

Proving this would be virtually identical to the proof of Theorem 3.3.7, except using $\theta\epsilon$ instead of $(1/2)\epsilon$ in the bounds leading up to (3.3.25), and $(1 - \theta)\epsilon$ instead of $\epsilon/2$ when we apply Theorem 3.3.3.

We now state a slightly technical corollary that we will use.

Corollary 3.3.9 *Let* $f(x) \in H^s$. *Let* $B_{X'}$, $B_{K'}$ *be as in Theorem 3.3.7. Then:*

$$\left\| \mathcal{P}_{B_{X'} \times B_{K'} \backslash \mathrm{HF}(K)} f(x) \right\|_{H^s} \leq \mathfrak{H}^s_+ (\tilde{g}(\vec{x})) \mathfrak{H}^{-s}_+ (e^{-x^2/\sigma^2})$$

$$\times \left(\|(1 - P_X(\vec{x})) f(\vec{x})\|_{H^s} + \epsilon \right) + \left\| \mathcal{P}_{\mathrm{HF}(K)} f(x) \right\|_{H^s} \tag{3.3.27}$$

Proof Repeat the proof of Theorem 3.3.7. However, instead of bounding $\left\|\mathcal{P}_{\mathrm{HF}(K)}f(x)\right\|_{H^s}$ using Theorem 3.3.3 to bound this term, we simply leave it as it is. $\square$

3.4 Phase Space Numerics

We now discuss algorithms for doing the aforementioned computations. More precisely, we describe an algorithm for computing the WFT coefficients of a function, at least for $\vec{a}$ restricted to a finite region. We also describe an algorithm for computing phase space projections.

The algorithm consists of computing the products $f(\vec{x})\tilde{g}(\vec{x} - \vec{a}x_0)$, followed by Fourier transforming the results. Due to the spatial decay of $\tilde{g}(\vec{x})$, we can truncate the domain to a reasonably small box surrounding $\vec{a}x_0$ with minimal error.

Algorithm 1 (Calculation of Windowed Fourier Transforms)
This algorithm calculates, for a function $f(x)$, the WFT coefficients $f_{(\vec{a},\vec{b})}$ for $\vec{a} \in A \subseteq \mathbb{Z}^N$. We assume that the frequencies are bounded above by $k_{\sup}$. The lattice spacing in frequency, k_0 is taken to be $2\pi/L_\epsilon$; taking it to be any larger than this yields only logarithmic improvements in computational complexity.

1. *Let $A \subset \mathbb{Z}^N$ be some set of position coordinates.*
2. *For each $\vec{a} \in A$ multiply $f(\vec{x})\tilde{g}(\vec{x} - \vec{a}x_0)$ for $\vec{x} \in [-L_\epsilon, L_\epsilon]^N + \vec{a}x_0$ only.*
3. *Calculate the Fast Fourier Transform of $f(\vec{x})\tilde{g}(\vec{x} - \vec{a}x_0)$ on this region. For each $\vec{a}$, the resulting function is $f_{(\vec{a},\vec{b})}$.*

We observe that this algorithm is local in space. This means that if `arange` is a finite region, then the computational cost is proportional to:

$$|A| \, k_{\sup}^N \left|\mathrm{supp}_\epsilon \, \tilde{g}(\vec{x})\right|^N \log(k_{\sup} |\mathrm{supp}\, \tilde{g}(\vec{x})|)$$

Here, $A \subset \mathbb{Z}^N$ is the region of space in which we want to compute the WFT coefficients.

The reason for this complexity is as follows. For each $\vec{a} \in A$, we need to compute an FFT. The FFT is computed on a region having size $\left|\mathrm{supp}_\epsilon\right|^N$, and the lattice spacing in this region is $2\pi/k_{\sup} = O(1/k_{\sup})$. Thus, there are $O(\left|\mathrm{supp}_\epsilon\right|^N k_{\sup}^N) = M$ data points in this region. The FFT has computational complexity $M \log(M)$. In addition, we need to compute $|A|$ of these FFT's.

Remark 3.4.1 As an example, consider the case when

$$A = \{\vec{a} \in \mathbb{Z}^N : L_{\mathrm{in}} \le |\vec{a}xs|_\infty \le L_{\mathrm{in}} + w\}$$

Then the size of A is proportional to:

$$\left| [-(L_{in} + w), (L_{in} + w)]^N \setminus [-L_{in}, L_{in}]^N \right| / x_0^N$$

If $L_{in} \gg w$, then this is of order L_{in}^{N-1}, and the computational complexity is $O(L_{in}^{N-1} k_{sup}^N \log(L_{in} k_{sup}))$.

We consider this case since this is what the TDPSF requires.

Phase space projections can also be computed. Let $F \subset \mathbb{Z}^N \times \mathbb{Z}^N$ be finite. Let $A = \{\vec{a} \in \mathbb{Z}^N : \exists \vec{b} \in \mathbb{Z}, (\vec{a}, \vec{b}) \in F\}$. Then we provide the phase space projection algorithm:

Algorithm 2 (Phase Space Projection Algorithm)

This algorithm computes the phase space projection onto a region of phase space F.

1. Compute $A = \{\vec{a} : (\vec{a}, \vec{b}) \in F\}$. Assume this is finite.
2. Compute $f_{(\vec{a}, \vec{b})}$ for $\vec{a} \in A$ as in Algorithm 1.
3. Define a new function $f^t : \mathbb{Z}^N \times \mathbb{Z}^N \to \mathbb{C}$ by:

$$f^t_{(\vec{a}, \vec{b})} = f_{(\vec{a}, \vec{b})}, \quad (\vec{a}, \vec{b}) \in F$$

$$f^t_{(\vec{a}, \vec{b})} = \quad 0, \quad (\vec{a}, \vec{b}) \notin F$$

4. Compute the inverse WFT of f^t. The result approximates $\mathcal{P}_F f(x)$, with errors caused by the truncation of $\tilde{g}(\vec{x})$ in Algorithm 1.

Clearly, the computational complexity of Algorithm 2 is of the same order as that of Algorithm 1.

Chapter 4
Description of Time Dependent Phase Space Filters

We now describe the TDPSF (Time Dependent Phase Space Filter) in more detail. We first begin with a motivating example, namely the case where we consider the semiclassical limit of (1.1.1).

4.1 An Introductory Example

Consider the following simple Schrödinger equation, with $V(x)$ a smooth, rapidly decaying potential.

$$\partial_t \Psi(\vec{x}, t) = (-\hbar^2(1/2)\Delta + V(x))\Psi(\vec{x}, t) \tag{4.1.1}$$

In the limit when $\hbar \to 0$, one can derive the following kinetic equation for $\rho(\vec{x}, t) = |\Psi(\vec{x}, t)|^2$:

$$\partial_t \tilde{\rho}(\vec{x}, \vec{k}, t) = (\vec{k} \cdot \nabla_x)\tilde{\rho}(\vec{x}, \vec{k}, t) + (\nabla V(\vec{k}) \cdot \nabla_k)\tilde{\rho}(\vec{x}, \vec{k}, t) \tag{4.1.2a}$$

$$\rho(\vec{x}, t) = \int \tilde{\rho}(\vec{x}, \vec{k}, t)d\vec{k} \tag{4.1.2b}$$

This equation is simple because it can be solved by the method of characteristics. The characteristic curve of (4.1.2) passing through the point $(\vec{x}, \vec{k})$ is the classical trajectory of a particle at the point $\vec{x}$ with initial velocity $\vec{k}$. Now, suppose that we are considering (4.1.2) on a box sufficiently large so that $V(x) \approx 0$ near the edge of the box.

In that case, near the boundary, the characteristic curve at $(\vec{x}, \vec{k})$ is parameterized locally by $(\vec{x} + \vec{k}t, \vec{k})$. Thus, it is easy to determine whether the flow is incoming or

© The Author(s), under exclusive license to Springer Nature Singapore Pte Ltd. 2023
A. Soffer et al., *Time Dependent Phase Space Filters*, SpringerBriefs on PDEs and Data Science, https://doi.org/10.1007/978-981-19-6818-1_4

outgoing near the boundary. We merely check whether $(\vec{x} + \vec{k}t, \vec{k})$ is moving in or out of the box. The algorithm is, therefore, as follows.

Surround the box $[-L_{in}, L_{in}]^N$ with an extra region (in the $\vec{x}$ direction) of width w. We let $L_{buff} = L_{in} + w$. We assume that the problem is such that the velocity is bounded above by k_{sup}. Then, inside the region $[-(L_{in} + w), (L_{in} + w)]^N \setminus [-L_{in}, L_{in}]^N$, we filter the outgoing trajectories every time $T_{st} = w/k_{sup}$. That is, letting $\tilde{\rho}(\vec{x}, \vec{k}, t)$ be the density, we set $\tilde{\rho}(\vec{x}, \vec{k}, t) = 0$ at the points $(\vec{x}, \vec{k})$ (with $\vec{x} \in [-(L_{in} + w), (L_{in} + w)]^N \setminus [-L_{in}, L_{in}]^N)$ where $(\vec{x} + t\vec{k}, \vec{k})$ is a trajectory which is leaving the box in the time interval $[0, T_{st}]$.

Thus, classical trajectories which are leaving the box are deleted before they reach the boundary, while trajectories which are not leaving the box are retained, and perfectly accurate propagation is achieved.

4.2   The TDPSF Algorithm

The TDPSF algorithm is an attempt to perform this procedure for (1.1.1). The primary sticking point is the Heisenberg uncertainty principle. We can no longer localize the solution precisely on outgoing positions and momenta.

However, by using a filter with good phase space localization, we can come close to extending this procedure to Schrödinger type equations. The only region of phase space where this works poorly is the region near $\vec{k} = 0$, due to the inability to localize a function only on outgoing trajectories.

Therefore, the algorithm we propose is as follows.

Suppose we have an initial condition $\Psi_0(x)$. The initial condition must be well localized in $[-(L_{in} + w), (L_{in} + w)]^N$, measured in H^s.

We decompose $\Psi_0(x, 0) = \sum_{j \in J} \Psi_{0j} \phi_j(x)$. We then split Ψ_0 up into framelets coming from the regions NECC $\cap$ BB, NECC^C and NECC $\cap$ BB^C.

We remove all framelets outside the set NECC $\cap$ BB.

It turns out that for a frame with good phase space localization, NECC and BAD are nearly mutually exclusive. This occurs because framelets, when propagated under the free flow, almost completely retain their coherence, and move either into the box or out of the box (but not both). Thus, by removing framelets outside NECC $\cap$ BB, we have removed nearly all of the outgoing waves.

Because of this, it is now most likely safe to propagate the remainder under the periodic flow, since the remainder consists of an initial condition that will not leave the box in the near future (with "near future" defined to be $[0, T_{st}]$).

The only time this is not true for the WFT is if a significant number of slow waves have reached the boundary. Every time T_{st}, we check if this has occurred. If so, we raise an exception.

When we reach time T_{st}, we go back to step one. That is, taking $\mathfrak{U}_b(T_{st})\Psi_{0,\text{modified}}$ as the new initial condition, we again filter off the outgoing waves. We repeat for as long as necessary. We assume that we can represent the function $\Psi(\vec{x}, t)$, restricted to the region $[-L_{\text{trunc}}, L_{\text{trunc}}]^N$ with periodic boundaries, with high accuracy. In our implementation, we store evenly spaced samples of $\Psi(\vec{x}, t)$, but other representations (e.g. finite element) can be used. By "high accuracy", we mean simply that the errors caused in this step are small in comparison to boundary error.

We further assume that $\mathfrak{U}_b(t)$ can be computed with accuracy very high relative to boundary error. The exact method of implementation is unimportant (provided it is sufficiently accurate): we use the FFT/Split Step propagation algorithm, but other methods (high order FDTD or finite elements) can be substituted.

The operators $\mathcal{P}_{\text{OUT}}$ and $\mathcal{P}_{\text{AMB}}$ are projections onto outgoing and ambiguous framelets, respectively. Their implementation is discussed in Sect. 3.4, in particular Algorithm 2. The number T_{st} is the time between filterings, which must be sufficiently small (see Sect. 7.4, in particular (7.4.1e) and (7.4.1g) for a precise formula).

Algorithm 4.2.1 (Propagation Algorithm)
This is the main propagation algorithm. In this algorithm, we consider σ, x_0, k_0 to be fixed. L_{in} and $T_{\mathfrak{M}}$ are also considered fixed. The initial data is considered fixed, and localized inside $[-L_{\text{in}}, L_{\text{in}}]^N$. The approximation will be denoted by $\Theta(x, t)$. Also, fix a small tolerance $\epsilon > 0$.

1. *Before beginning, precalculate the set of framelets which are outgoing, and those which are ambiguous.*
2. *Define $\Theta(x, t)$ iteratively as follows. Loop over $n = 0 \ldots T_{\mathfrak{M}}/T_{st}$. In what follows, the propagator $\mathfrak{U}(t)$ is calculated by Algorithm 4.3.2.*

   (a) For $t \in [(n-1)T_{st}, nT_{st})$, define

   $$\Psi(x, t) = \mathfrak{U}(t - (n-1)T_{st})\Psi(x, (n-1)T_{st})$$

   (b) For $t = nT_{st}$, define:

   $$\Theta(x, nT_{st}) = (1 - \mathcal{P}_{\text{NECC}^c})\mathfrak{U}(T_{st})\Psi(x, (n-1)T_{st})$$

   (c) At each integer multiple of T_{st}, compute

   $$\|\mathcal{P}_{\text{BAD}\cap\text{NECC}}\mathfrak{U}(T_{st})\Psi(x, (n-1)T_{st})\|_{H^s}$$

   where $\text{BAD} \cap \text{NECC}$ is the set of ambiguous framelets. If this quantity is greater than ϵ, stop the program and notify the user.

Note that it is far more efficient to compute $(1 - \mathcal{P}_{\text{NECC}^c})$ than it would be to compute $\mathcal{P}_{\text{NECC}}$. This is the case because the necessary framelets are mostly localized in the region $[-L_{\text{trunc}}, L_{\text{trunc}}]^N \setminus [-L_{\text{in}}, L_{\text{in}}]^N$, implying that there are

$O(L_{\text{in}}^{N-1} w k_{\text{sup}}^N)$ such framelets. However, there are $O(L_{\text{in}}^N k_{\text{sup}}^N)$ framelets in NECC, and in the regime $L_{\text{in}} \gg w$ (which we are considering) this is a substantial difference.

4.3 Implementation of the Algorithm

The algorithm we have described is, to a great extend, independent of the particular method of implementation. However, we sketch out one possible method of implementing it here, namely the FFT/Split Step algorithm.

We fix a grid spacing δx satisfying $\pi/\delta x > k_{\text{sup}}$, and timestep δt. This corresponds to a lattice spacing in momentum of $2\pi/L_{\text{trunc}}$, with maximal momentum $\pi/\delta x$. In practice, one usually oversamples, taking instead $3\pi/2\delta x > k_{\text{sup}}$, to allow detection of spectral aliasing.

We then use the Fast Fourier transform algorithm to implement the standard split step/Trotter-Kato formula spectral propagator:

Algorithm 4.3.2 (Split Step Propagation Algorithm)
This algorithm approximates the propagation of (1.1.1) *on the region* $[-L_{\text{trunc}}, L_{\text{trunc}}]^N$. *The accuracy is* $O(\delta t^2)$ *time and* $O(\delta x^\omega)$ *("spectral accuracy") in space. It is* $O(\delta t^3)$ *in time if* (1.1.1) *is linear.*

First, the discrete operator $e^{i(1/2)\Delta t} f(x)$ *is defined by computing the FFT of* $f(x)$, *multiplying by* $e^{ik^2 t}$ *and then computing the inverse FFT of the result.*

With this in mind, the algorithm is as follows (taking $\Psi_0(x)$ *as initial condition, and assuming t is a multiple of* δt):

1. *Define* $\Psi_{1/2}(x) = e^{i(1/2)\Delta \delta t/2} \Psi_0(x)$.
2. *For* $j = 0, \ldots, t/\delta t - 2$, *define:*

$$\Psi_{j+1+1/2}(x) = e^{i(1/2)\Delta\delta t} e^{-ig(j\delta t, \vec{x}, \cdot)\Psi_{j+1/2}(x)\delta t} \Psi_{j+1/2}(x)$$

3. *Finally, define* $\Psi_{t/\delta t}(x) \approx \Psi(x, t)$ *by:*

$$\Psi_{t/\delta t}(x) = e^{i(1/2)\Delta(\delta t/2)} e^{-ig(j\delta t, \vec{x}, \cdot)\Psi_{j+1/2}(x)\delta t} e^{i(1/2)\Delta\delta t} \Psi_{t/\delta t-3/2}(x)$$

This is a numerical realization of the Trotter-Kato product formula:

$$\mathfrak{U}(t) \approx e^{i(1/2)\Delta(-\delta t/2)} \left[\prod_{j=1}^{t/\delta t} e^{i(1/2)\Delta\delta t} e^{-ig(j\delta t, \vec{x}, \cdot)\Psi(x, j\delta t)\delta t} \right] e^{i(1/2)\Delta\delta t/2} \quad (4.3.1)$$

Algorithm 4.3.2 is "spectrally accurate" in x, of order $O(\delta t^2)$ in time (for nonlinear problems, for linear problems it increases to $O(\delta t^3)$), and has cost

$O(M^N \ln M)$ per timestep (where M is the number of data points in the grid, per dimension). For this reason it is a popular method of propagating dispersive waves.

4.4 An Intuition of the Algorithm

The framelets in in NECC^C consist of framelets which are moving out of the box under the free flow $e^{i(1/2)\Delta t}$. Thus, there is little error caused by removing them.

For the WFT frame, the framelets in $\text{NECC} \cap \text{BB}^C$ consist of framelets which are outside the box, but are moving inward under the free flow. If the initial condition $\Psi(x, 0)$ is well localized, the only way such framelets can exist is if waves moved out of the box, turned around and came back. This is extremely unlikely. Thus, there is little error caused by removing these framelets.

The remainder consist of framelets in $\text{NECC} \cap \text{BB} \cap \text{BAD}$. In general, little can be said about these framelets. But for the WFT, these consist of framelets which are moving slowly, more slowly than a certain velocity k_{inf}. We make this term small merely by assuming it to be true. In practice, it may not be, although we outline (non-rigorously) methods of dealing with this.

We now consider the remaining framelets. Apart from the slowly moving ones, the framelets in $\text{NECC} \cap \text{BB}$ are not coming close to the boundaries of $[-L_{\text{trunc}}, L_{\text{trunc}}]^N$. Thus, the boundary conditions we have chosen (periodic, in this case) are irrelevant. This is true for a short time, say a time T_{st}.

In the event that the slowly moving framelets in $\text{BAD} \cap \text{NECC}$ do reach the boundary, then an exception is raised.

4.5 Possible Enhancements

One obvious improvement to our algorithm is useful for dealing with Hamiltonians of the form $H = -(1/2)\Delta + V(x) + f(|\Psi(\vec{x}, t)|)$ with $V(x)$ a localized potential (possibly of long range type). Instead of trying to determine whether the free trajectory of a given framelet, namely $\vec{a}x_0 + \vec{b}k_0t$ is leaving the box sufficiently fast, we try to determine whether the interacting trajectory $\gamma(\vec{a}x_0, \vec{b}k_0, t)$ is leaving the box. The interacting trajectory is the trajectory obeyed by a classical particle with velocity $\vec{b}k_0$ moving in the potential $V(x)$. Intuitively this is the right thing to do, although we do not know how to prove this.

The main unknown factor in our algorithm is k_{inf}, the smallest relevant frequency. The problem will appear as follows. Suppose that at some time NT_{st}, we find that the mass sitting on the framelets in $\text{BAD} \cap \text{NECC}$ is not small. We can reduce k_{inf} by increasing σ, the width of the framelets in x. The cost of doing this is that it becomes necessary to increase the width of the buffer region w to $w = O(k_{\text{inf}}^{-1})$. Therefore the computational complexity of the TDPSF algorithm grows like $O(k_{\text{inf}}^{-1})$.

4.6   Slow Waves Multiscale Resolution

The problem of slow waves is the weakest point of the algorithm described here, and is in fact a weak point of most methods. Absorbing potentials reflect nearly all low frequency waves, and Dirichlet-to-Neumann boundaries are usually implemented based on high frequency approximations which also break down near $k = 0$. The basis of the problem is the Heisenberg uncertainty principle; to resolve waves with resolution k_{inf} in the frequency domain, one needs a spatial domain at least as large as k_{inf}^{-1}.

In [63] we construct an extension of the TDPSF which treats the problem in time proportional to $\log k_{\text{inf}}$ rather than k_{inf}^{-1}. The basic idea is as follows. Instead of solving the problem on a uniformly sampled grid on $[-L_{\text{in}}, L_{\text{in}}]^N$, we extend the computational domain from $[-L_{\text{in}}, L_{\text{in}}]$ to $[-2L_{\text{in}}, 2L_{\text{in}}]$. However, on the region $[-2L_{\text{in}}, -L_{\text{in}}] \cup [L_{\text{in}}, 2L_{\text{in}}]$, we sample the grid at a rate $2\delta x$ instead of δx. We apply a TDPSF filter, as described in this work on the boundary of the region $[-L_{\text{in}}, L_{\text{in}}]$, but letting waves with frequency k_{inf} propagate to the next grid. On the boundary of the second grid, we filter waves with frequencies below $k_{\text{inf}}/2$ on a filter region twice as large as that on the first grid. This process can be repeated M times to filter waves below $2^{-M} k_{\text{inf}}$. However, since we reduce the sampling rate each time we extend the grid, the computational cost is proportional to M, not 2^M, thus we achieve $O(\log k_{\text{inf}})$ time complexity.

The effectiveness of this algorithm is the fact that the NLSE is being solved on a non-rectangular region of phase space, namely a covering of $\{(x, k) : |k| \le |x|^{-1}\}$ by rectangles. Since the TDPSF is a phase space based algorithm, it is very simple to extend it to such regions. It is not even obvious what it means to extend the PML, absorbing potential or Dirichlet-to-Neumann boundary to such a computational domain.

Chapter 5
A More Practical Discussion: How to Choose the Parameters

In this Chapter, we provide a more practical and friendly description of choosing the parameters in the method, for general wave equations. As the description is important for practical purposes, we put it in a separate chapter.

- **Step 1:** In order to start the program, one should choose an initial good guess of the parameters of the computation. For this purpose, one needs to identify the equation to be solved, in particular the dispersion relation $\omega(k)$ and the size of the domain of interest, $[-L_{\text{in}}, L_{\text{in}}]^N$, where N is the dimension. An expression for the group velocity is given by $v = \nabla_k \omega(k)$ where k is the wave number. We then compute the following two crucial numbers: the window in k space where the solution (is expected to) live: k_{inf} and k_{sup}.
- **Step 2:** We choose the lattice spacing to be $d_x = 1/3k_{sup}$ and we need to ensure that the domain size $L_{in} >> 1/k_{inf}$. Next, we choose the time step d_t to ensure accuracy and stability as well as the time T for which the solution is desired (T is large in general). We also fix the accuracy ϵ desired.
- **Step 3:** In this step, we begin to determine the filter parameters. The mother Gaussian for the frames has parameter σ which is the variance. The frame parameters are the steps in x and v directions: the velocity direction step $k_0 = (1/3)k_{inf}$, but is should be strictly smaller than v_{min}. Then, choose x_0 such that $x_0 k_0 = (2\pi)/4$. A division by $2n > 4$ improves the accuracy.
- **Step 4:** Next, we find the size of the filter domain w. The Filter domain is the region of thickness w around the boundary. First, we fix the filter timing d_T: the time intervals at which filtering is done. It should be at least 20 times d_t : $d_T > 20d_t$. Then we choose $w > 3v_{max}d_T$. Finally choose σ such that $w > n\sigma$; $n > 6$, and also insure that $x_0 << \sigma$.
- **Step 5:** The frame is generated by translation of a mother frame, and we take it to be a Gaussian, with variance σ. The dual frame in this case is also generated by one mother function, and the formulas and properties are described in the detail in the relevant sections. See for example formula 3.2.11.

© The Author(s), under exclusive license to Springer Nature Singapore Pte Ltd. 2023
A. Soffer et al., *Time Dependent Phase Space Filters*, SpringerBriefs on PDEs and Data Science, https://doi.org/10.1007/978-981-19-6818-1_5

Then, each frame is translation of the mother one by a step x_0 in space, and a step of size k_0 in the phase. We must choose $x_0 k_0 < 2\pi$. To have accurate velocity of the wave, we choose $k_0 = k_{inf}/3$. Here we equate k with velocity (the case when $\omega(k) = k^2/2$). Recalling that k_{inf} is the lowest expected frequency.

- **Step 6:** To minimize the error of the filtering, we need to take $k_{inf} > 3/\sigma$. (in this case the error from removing a Gaussian with minimum velocity, will be of the order Ae^{-3^2}, A is the amplitude of the removed gaussian. If you double the size of σ, the error will be Ae^{-6^2}. From here, one could see that that removing a gaussian with small velocity is producing big error. But the program can store/leave/notify when such gaussian shows up.

- **Step 7:** The expansion for a function in the filter domain, is done by taking the usual scalar product of the function with the dual frame functions. The width of the filter is determined by the max velocity, mainly. It should be wide enough so that its length w, should be at least $3D_T$. D_T is the time interval after which we apply the filtering. It should be chosen as long as possible, to reduce the number of filtering steps. But of course, too long, will make w too big. w should also be much bigger than the main support of the mother frame. When a Gaussian is too slow mover, and higher accuracy is needed, one can remove such Gaussians and upload on a new computational domain, with size twice as large as the original one, but with the same number of lattice point as the old one. In the linear case all such Gaussians can be propagated under the original flow independently of the rest of the solution.

Chapter 6
The Behavior of Gaussian Framelets Under the Free Flow

In this section we study the behavior of Gaussian framelets under the free flow, $e^{i(1/2)\Delta t}$. This is quite explicit, because we can write $e^{i(1/2)\Delta t}\phi_{(\vec{a},\vec{b})}(\vec{x})$ in closed form:

$$e^{i(1/2)\Delta t}\phi_{(\vec{a},\vec{b})}(\vec{x}) = e^{i(1/2)\Delta t}\pi^{-N/4}\sigma^{-N/2}e^{ik_0\vec{b}\cdot\vec{x}}e^{-|\vec{x}-\vec{a}x_0|_2^2/2\sigma^2}$$

$$= \frac{\exp(i\vec{b}k_0\cdot(\vec{x}-\vec{b}k_0/2t-\vec{a}x_0))}{\pi^{N/4}\sigma^{N/2}(1+it/\sigma^2)^{N/2}}\exp\left(\frac{-|\vec{x}-\vec{b}k_0t-\vec{a}x_0|_2^2}{2\sigma^2(1+it/\sigma^2)}\right) \quad (6.0.1)$$

This allows us to compute precisely most of our framelet functions (error, relevance, etc).

We begin with a general result, which allows us to control the error associated with approximating Fourier multipliers on $\mathbb{R}^N$ by restricting them to a box. This result is sufficiently general to allow for the use of certain kinds of low pass filters (in frequency) on the box, although we do not use it in this generality.

Theorem 6.0.1 *Let $S(i\nabla)\varphi(\vec{x})$ satisfy the hypothesis of the Poisson summation formula, that is $|S(i\nabla)\varphi(x)| \le C\langle x\rangle^{N+\epsilon}$ and $\left|S(i\vec{k})\hat{\varphi}(k)\right| \le C\langle k\rangle^{N+\epsilon}$. Let $S(\vec{k})$, $S_b(\vec{k})$ be continuous bounded Fourier multiplication operators which are equal for $\vec{k} \in B$ (where B is some closed set).*
Then:

$$\left\|S(i\nabla)\varphi(\vec{x}) - \sum_{\vec{k}\in B}e^{i\pi\vec{k}\cdot\vec{x}/L}S_b(\pi\vec{k}/L)\hat{\varphi}(\pi\vec{k}/L)\right\|_{H_b^s}$$

$$\le \|S(i\nabla)\varphi(\vec{x}+2L\vec{n})\|_{H^s(([-L,L]^N)^C)}$$

$$+ \left\|\hat{\varphi}(\vec{k})\right\|_{H^s(B^C)}\sup_{\vec{k}\in B^C}\left|S(\vec{k})-S_b(\vec{k})\right| \quad (6.0.2)$$

© The Author(s), under exclusive license to Springer Nature Singapore Pte Ltd. 2023
A. Soffer et al., *Time Dependent Phase Space Filters*, SpringerBriefs on PDEs and Data Science, https://doi.org/10.1007/978-981-19-6818-1_6

Remark 6.0.2 A result which is similar to this one appeared in [57], where it is used to show that systems on a large enough box can exhibit transient radiative behavior (over short times) in the same way that systems on $\mathbb{R}^N$ can.

Remark 6.0.3 We only use this theorem with $S(i\nabla) = S_b(i\nabla)$; thus, the last term in (6.0.2) is zero for our purposes . The more general version might be useful when studying the effects of low pass filters on numerical schemes. For many years (since, e.g. [54]), low pass filters have been applied to numerical schemes in order to preserve numerical stability. This result might be useful in proving error bounds for such schemes.

Proof The Poisson summation formula states that:

$$\sum_{n\in\mathbb{Z}^d} f(\vec{x} + n2L) = \sum_{k\in\mathbb{Z}^d} e^{i\pi\vec{k}\cdot\vec{x}/L}\,\hat{f}\,(\pi k/L) \tag{6.0.3}$$

We let $\hat{f}(\vec{k}) = S(\vec{k})\hat{\varphi}(\vec{k})$. Then, by rearranging (6.0.3), we find:

$$S(i\nabla)\varphi(\vec{x}) - \sum_{\vec{k}\in\mathbb{Z}^N} e^{i\pi\vec{k}\cdot\vec{x}/L}S(\pi\vec{k}/L)\hat{\varphi}(\pi\vec{k}/L) = -\sum_{\substack{\vec{n}\in\mathbb{Z}^N \\ \vec{n}\neq 0}} S(i\nabla)\varphi(\vec{x} + 2L\vec{n})$$

$$\tag{6.0.4}$$

Now, we observe that $S(\vec{k})$ and $S_b(\vec{k})$ are equal on B. We add and subtract

$$\sum_{\pi\vec{k}/L\in\mathbb{Z}^N} e^{i\pi\vec{k}\cdot\vec{x}/L}(S_b(\pi\vec{k}/L) - S(\pi\vec{k}/L))\hat{\varphi}(\pi\vec{k}/L)$$

to both sides of (6.0.4), to obtain:

$$S(i\nabla)\varphi(\vec{x}) - \sum_{\vec{k}\in\mathbb{Z}^N} e^{i\pi\vec{k}\cdot\vec{x}/L} S_b(\pi\vec{k}/L)\hat{\varphi}(\pi\vec{k}/L)$$

$$= -\sum_{\substack{\vec{n}\in\mathbb{Z}^N \\ \vec{n}\neq 0}} S(i\nabla)\varphi(\vec{x}+2L\vec{n}) + \sum_{\pi\vec{k}/L\in\mathbb{Z}^N} e^{i\pi\vec{k}\cdot\vec{x}/L}(S_b(\pi\vec{k}/L) - S(\pi\vec{k}/L))\hat{\varphi}(\pi\vec{k}/L)$$

We again apply (6.0.3), and observe that:

$$\sum_{\pi\vec{k}/L\in\mathbb{Z}^N} e^{i\pi\vec{k}\cdot\vec{x}/L}(S_b(\pi\vec{k}/L) - S(\pi\vec{k}/L))\hat{\varphi}(\pi\vec{k}/L)$$

$$= \sum_{\vec{n}\in\mathbb{Z}^N} (S_b(i\nabla) - S(i\nabla))\varphi(\vec{x} + 2L\vec{n})$$

We now take norms and apply the triangle inequality. We find that:

$$\sum_{\substack{\vec{n}\in\mathbb{Z}^N \\ \vec{n}\neq 0}} \|S(i\nabla)\varphi(\vec{x}+2L\vec{n})\|_{H^s_b} = \|S(i\nabla)\varphi(\vec{x}+2L\vec{n})\|_{H^s(([-L,L]^N)^C)}$$

and that:

$$\sum_{\vec{n}\in\mathbb{Z}^N} \|(S_b(i\nabla)-S(i\nabla))\varphi(\vec{x}+2L\vec{n})\|_{H^s_b} = \|(S_b(i\nabla)-S(i\nabla))\varphi(\vec{x}+2L\vec{n})\|_{H^s}$$

$$\leq \left\|\hat{\varphi}(\vec{k})\right\|_{H^s(B^C)} \sup_{\vec{k}\in B^C} \left|S(\vec{k})-S_b(\vec{k})\right|$$

We put everything together to obtain the result we seek.   □

We also have an alternate version of this result which works for arbitrary domains (provided Green's theorem applies), and linear Schrödinger operators of the form $H = -(1/2)\Delta + V(x)$. However, the error bound is only in L^2.

Theorem 6.0.4 *Let Ω be a bounded subset of $\mathbb{R}^N$ to which Green's theorem applies, and let $\Psi_0(x)$ be supported on Ω. Let $H = -(1/2)\Delta + V(x)$ for $V(x)$ real valued and localized, and let $H_b = -(1/2)\Delta_b + V(x)$ where $(1/2)\Delta_b$ is the Laplacian on Ω with Dirichlet boundaries. Then:*

$$\left\|[e^{iHt}-e^{iH_bt}]\Psi_0(x)\right\|_{L^2(\Omega)} \leq \left\|e^{iHt}\Psi_0(x)\right\|_{L^2(\Omega^C)} \tag{6.0.5}$$

Proof Let $e(x,t) = \chi_\Omega(x)[e^{iHt}-e^{iH_bt}]\Psi_0(x)$ be the error. Then $e(x,t)$ solves the Schrödinger equation on Ω with zero initial condition and boundary conditions given by $[e^{iHt}\Psi_0](x) - [e^{iH_bt}\Psi_0](x) = [e^{iHt}\Psi_0](x) - 0$ (for $x \in \partial\Omega$). We then observe that:

$$\partial_t \int_\Omega |e(x,t)|^2\,dx = 2\Re\int_\Omega \bar{e}(x,t)\partial_t e(x,t)dx = 2\Re\int_\Omega \bar{e}(x,t)(-iH)e(x,t)dx$$

$$= 2\Im\int_\Omega \bar{e}(x,t)(1/2)\Delta e(x,t) + 2\Im\int_\Omega \bar{e}(x,t)V(x)e(x,t)dx$$

$$= \left[\Im\int_\Omega \nabla\bar{e}(x,t)\cdot\nabla e(x,t)dx + \Im\int_{\partial\Omega} \bar{e}(x,t)\partial_{\vec{n}}e(x,t)dx\right] + 0$$

$$= \Im\int_{\partial\Omega} \bar{e}(x,t)\partial_{\vec{n}}e(x,t)dx$$

The last term is the probability flux through $\partial\Omega$. The potential term vanished since $V(x)$ is real. Thus

$$\partial_t \|e(x,t)\|^2_{L^2(\Omega)} = \text{flux}$$

and

$$\|e(x,t)\|^2_{L^2(\Omega)} \le \int_0^t \text{flux}(t)dt = \left\|e^{iHt}\Psi_0(x)\right\|^2_{L^2(\Omega^C)}$$

which is what we wanted to prove.   □

6.1   Error Estimates

Using Theorem 6.0.1 and Eq. (6.0.1), we can compute per-framelet error bounds in $L^2(\mathbb{R})$. Before we continue, we define a function we will use a number of times.

Definition 6.1.1 We define the Hermite Error Function, for x, k real and $s > 0$ to be:

$$\text{Herf}^s(x,k) = \frac{2}{\sqrt{\pi}} \int_0^x \left(\partial_w^s e^{iwk} e^{-w^2/2}\right)\left(\partial_w^s e^{-iwk} e^{-w^2/2}\right) dw \qquad (6.1.1)$$

Note that $\text{Herf}^0(x,k) = \text{erf}(x)$. We also define $\text{Herfi}^s(x,k)$ to be the inverse function of $\text{Herf}^s(\,\cdot\,,k)$.

Remark 6.1.2 We observe that to leading order in k (as k becomes large), that

$$\text{Herf}^s(x,k) = |k|^{2s}\,\text{erf}(x) + O(|k|^{2s-1})$$

In $L^2 = H^0$, $\text{Herf}^s(x,k) = \text{erf}(x)$. In higher Sobolev spaces, they can be determined by a symbolic computation utility, e.g. Maple.

We will use the Herf^s function when we need to compute the L^2 norm of derivatives of gaussians.

Proposition 6.1.3 *In H^s, we can compute the framelet functionals:*

$$\mathcal{R}^s_{(\vec{a},\vec{b})}(t)^2 = \mathcal{R}^0_{(\vec{a},\vec{b})}(t)^2 + 2^{-N}(\sigma^{-1}(1+t^2/\sigma^4)^{1/2})^{2s}$$

$$\times \sum_{j=1}^N \left(\left[\text{Herf}^s\left(\frac{L_{\text{in}}+\vec{b}_j k_0 t + \vec{a}_j x_0}{\sigma\sqrt{1+t^2/\sigma^4}}, \vec{b}_j k_0(\sigma^{-1}(1+t^2/\sigma^4)^{-1/2})\right)\right.\right.$$

$$\left.- \text{Herf}^s\left(\frac{-L_{\text{in}}+\vec{b}_j k_0 t + \vec{a}_j x_0}{\sigma\sqrt{1+t^2/\sigma^4}}, \vec{b}_j k_0(\sigma^{-1}(1+t^2/\sigma^4)^{-1/2})\right)\right]$$

$$\times \prod_{\substack{k=1\\k\ne j}}^N \left[\text{erf}\left(\frac{L_{\text{in}}+\vec{b}_k k_0 t + \vec{a}_k x_0}{\sigma\sqrt{1+t^2/\sigma^4}}\right) - \text{erf}\left(\frac{-L_{\text{in}}+\vec{b}_k k_0 t + \vec{a}_j x_0}{\sigma\sqrt{1+t^2/\sigma^4}}\right)\right]\right)$$

$$(6.1.2a)$$

$$\mathfrak{E}^s_{(\vec{a},\vec{b})}(t)^2 = \mathfrak{E}^0_{(\vec{a},\vec{b})}(t)^2 + (\mathcal{M}^s_{(\vec{a},\vec{b})})^2 - 2^{-N}(\sigma^{-1}(1+t^2/\sigma^4)^{1/2})^{2s}$$

$$\times \sum_{j=1}^{N} \left(\left[\mathrm{Herf}^s \left(\frac{L_{\mathrm{buff}} + \vec{b}_j k_0 t + \vec{a}_j x_0}{\sigma\sqrt{1+t^2/\sigma^4}}, \vec{b}_j k_0 (\sigma^{-1}(1+t^2/\sigma^4)^{-1/2}) \right) \right. \right.$$

$$\left. - \mathrm{Herf}^s \left(\frac{-L_{\mathrm{buff}} + \vec{b}_j k_0 t + \vec{a}_j x_0}{\sigma\sqrt{1+t^2/\sigma^4}}, \vec{b}_j k_0 (\sigma^{-1}(1+t^2/\sigma^4)^{-1/2}) \right) \right]$$

$$\times \prod_{\substack{k=1\\k\neq j}}^{N} \left[\mathrm{erf} \left(\frac{L_{\mathrm{buff}} + \vec{b}_k k_0 t + \vec{a}_k x_0}{\sigma\sqrt{1+t^2/\sigma^4}} \right) - \mathrm{erf} \left(\frac{-L_{\mathrm{buff}} + \vec{b}_k k_0 t + \vec{a}_k x_0}{\sigma\sqrt{1+t^2/\sigma^4}} \right) \right] \right)$$

$$(6.1.2b)$$

Proof By Theorem 6.0.1, to calculate $\mathfrak{E}^s_{(\vec{a},\vec{b})}(t)$, we need only compute the mass outside the box $B = [-(L_{\mathrm{in}}+w),(L_{\mathrm{in}}+w)]^N$. We observe that $\left\| \phi_{(\vec{a},\vec{b})}(\vec{x}) \right\|_{H^s} = 1 + \mathcal{M}^s_{(\vec{a},\vec{b})}$, so therefore:

$$\left\| e^{i(1/2)\Delta t} \phi_{(\vec{a},\vec{b})}(\vec{x}) \right\|_{H^s(\mathbb{R}^N\setminus[-(L_{\mathrm{in}}+w),(L_{\mathrm{in}}+w)]^N)}$$

$$= 1 + \mathcal{M}^s_{(\vec{a},\vec{b})} - \left\| e^{i(1/2)\Delta t} \phi_{(\vec{a},\vec{b})}(\vec{x}) \right\|_{H^s([-(L_{\mathrm{in}}+w),(L_{\mathrm{in}}+w)]^N)}$$

We need to compute

$$\left\| \partial^s_{x_j} e^{i(1/2)\Delta t} \phi_{(\vec{a},\vec{b})}(\vec{x}) \right\|_{L^2([-L_{\mathrm{in}},L_{\mathrm{in}}]^N)}$$

for $j = 1 \ldots N$, and also for $s = 0$. We compute as follows:

$$\left\| \partial^s_{x_j} e^{i(1/2)\Delta t} \phi_{(\vec{a},\vec{b})}(\vec{x}) \right\|^2_{L^2([-L_{\mathrm{in}},L_{\mathrm{in}}]^N)} = \int_{[-L_{\mathrm{in}},L_{\mathrm{in}}]^N}$$

$$\left| \partial^s_{x_j} \frac{\exp(i\vec{b}k_0 \cdot (\vec{x} - \vec{b}k_0/2t - \vec{a}x_0))}{\pi^{N/4}\sigma^{N/2}(1+it/\sigma^2)^{N/2}} \exp\left(\frac{-|\vec{x} - \vec{b}k_0 t - \vec{a}x_0|^2_2}{2\sigma^2(1+it/\sigma^2)} \right) \right|^2 d\vec{x}$$

$$= \frac{1}{\pi^{N/2}\sigma^N\sqrt{1+t^2/\sigma^4}} \int \left| \partial^s_{x_j} e^{i\vec{b}k_0 \cdot \vec{x}} \exp\left(\frac{-|\vec{x} - \vec{b}k_0 t - \vec{a}x_0|^2}{2\sigma^2(1+it\sigma^2)} \right) \right|^2 d\vec{x}$$

$$(6.1.3)$$

We change variables to $\vec{y}_j = \sigma^{-1}(1 + t^2/\sigma^4)^{-1/2}(\vec{x}_j - \vec{b}_j k_0 t - \vec{a}_j x_0)$, then $\sigma\sqrt{1 + t^2/\sigma^4} d\vec{y}_j = d\vec{x}_j$.

$$(6.1.3) = \left[\prod_{\substack{1 \le k \le N \\ k \ne j}} \int_{(-L_{in}+\vec{b}_j k_0 t+\vec{a}_j x_0)/\sigma\sqrt{1+t^2/\sigma^4}}^{(L_{in}+\vec{b}_j k_0 t+\vec{a}_j x_0)/\sigma\sqrt{1+t^2/\sigma^4}} e^{-\vec{y}_j^2} d\vec{y}_j \right]$$

$$\left[\int_{\sigma^{-1}(1+t^2/\sigma^4)^{-1/2}(-L_{in}-\vec{b}k_0 t-\vec{a}x_0)}^{\sigma^{-1}(1+t^2/\sigma^4)^{-1/2}(L_{in}-\vec{b}k_0 t-\vec{a}x_0)} \right.$$

$$\left. \left| (\sigma^{-1}(1 + t^2/\sigma^4)^{-s/2}) \partial_{y_j}^s e^{i\vec{b}_j k_0 \sigma^{+1}(1+t^2/\sigma^4)^{1/2}\vec{y}_j} e^{-y_j^2/2} \right|^2 dx_j \right] \qquad (6.1.4)$$

Evaluating the integrals yields:

$$(6.1.4) = (\sigma^{-1}(1 + t^2/\sigma^4)^{1/2})^{2s}$$

$$\times \left(\text{Herf}^s \left(\frac{L_{in} + \vec{b}_j k_0 t + \vec{a}_j x_0}{\sigma\sqrt{1+t^2/\sigma^4}}, \vec{b}_j k_0(\sigma^{-1}(1+t^2/\sigma^4)^{-1/2}) \right) \right.$$

$$\left. - \text{Herf}^s \left(\frac{L_{in} + \vec{b}_j k_0 t + \vec{a}_j x_0}{\sigma\sqrt{1+t^2/\sigma^4}}, \vec{b}_j k_0(\sigma^{-1}(1+t^2/\sigma^4)^{-1/2}) \right) \right)$$

$$2^{-N} \prod_{\substack{1 \le i \le N \\ i \ne j}} \left[\text{erf}\left(\frac{L_{in} + \vec{b}_i k_0 t + \vec{a}_i x_0}{\sigma\sqrt{1+t^2/\sigma^4}} \right) - \text{erf}\left(\frac{-L_{in} + \vec{b}_i k_0 t + \vec{a}_i x_0}{\sigma\sqrt{1+t^2/\sigma^4}} \right) \right]$$

$$(6.1.5)$$

We add this up for $j = 1..N$ (since we take derivatives in each component of $\vec{x}$) and add a term with $s = 0$. This yields the result we seek. A similar computation allows us to compute $\mathfrak{E}^s_{(\vec{a},\vec{b})}(t)$.   □

Remark 6.1.4 For the specific cases of L^2 and H^1, we include simpler formulas. We single out these cases because they are sufficient to encompass most simulations of practical interest.

In L^2, we obtain:

$$\mathfrak{E}^0_{(\vec{a},\vec{b})}(t) = 1 - 2^{-N/2} \prod_{j=1}^N \left[\text{erf}\left(\frac{(L + w) + \vec{b}_j k_0 t + \vec{a}_j x_0}{\sigma\sqrt{1+t^2/\sigma^4}} \right) - \right.$$

$$\left. \text{erf}\left(\frac{-(L + w) + \vec{b}_j k_0 t + \vec{a}_j x_0}{\sigma\sqrt{1+t^2/\sigma^4}} \right) \right]^{1/2} \qquad (6.1.6a)$$

$$\mathcal{R}^0_{(\vec{a},\vec{b})}(t)$$

$$= 2^{-N/2} \prod_{j=1}^{N} \left[\mathrm{erf}\left(\frac{L + \vec{b}_j k_0 t + \vec{a}_j x_0}{\sigma\sqrt{1 + t^2/\sigma^4}} \right) - \mathrm{erf}\left(\frac{-L + \vec{b}_j k_0 t + \vec{a}_j x_0}{\sigma\sqrt{1 + t^2/\sigma^4}} \right) \right]^{1/2}$$

$$(6.1.6b)$$

In $H^1(\mathbb{R})$, we find that $\mathcal{R}^1_{(\vec{a},\vec{b})}(t)$ is given by:

$$\left| \mathcal{R}^1_{(\vec{a},\vec{b})}(t) \right|^2 = \left| \mathcal{R}^0_{(\vec{a},\vec{b})}(t) \right|^2$$

$$+ 2^{-N} \sum_{k=1}^{N} \left\{ \left(\frac{(2bk_0 t - 2L_{\mathrm{in}} - 2ax_0)}{8\sqrt{\pi(1+t^2)}} \right. \right.$$

$$\times \left[e^{-(L_{\mathrm{in}}+\vec{a}_k x_0 + \vec{b}_k k_0 t)^2/(1+t^2)} - e^{-(L_{\mathrm{in}} - \vec{a}_k x_0 - \vec{b}_k k_0 t)^2/(1+t^2)} \right]$$

$$+ 8^{-1} \left(1 + 2b^2 k_0^2 \right) \left[\mathrm{erf}\left(\frac{Lb + ax_0 + bk_0 t}{\sqrt{1+t^2}} \right) + \mathrm{erf}\left(\frac{Lb - ax_0 - bk_0 t}{\sqrt{1+t^2}} \right) \right] \right)$$

$$\times \prod_{j=1, j\neq k}^{N} \left[\mathrm{erf}\left(\frac{L_{\mathrm{in}} + \vec{b}_j k_0 t + \vec{a}_j x_0}{\sigma\sqrt{1 + t^2/\sigma^4}} \right) - \mathrm{erf}\left(\frac{-L_{\mathrm{in}} + \vec{b}_j k_0 t + \vec{a}_j x_0}{\sigma\sqrt{1 + t^2/\sigma^4}} \right) \right] \right\}$$

$$(6.1.7a)$$

$$\left| \mathcal{E}^1_{(\vec{a},\vec{b})}(t) \right|^2 = \left| \mathcal{E}^0_{(\vec{a},\vec{b})}(t) \right|^2 + (\mathcal{M}^s_{(\vec{a},\vec{b})})^2$$

$$- 2^{-N} \sum_{k=1}^{N} \left\{ \left(\frac{(2bk_0 t - 2(L_{\mathrm{in}} + w) - 2ax_0)}{8\sqrt{\pi(1+t^2)}} \right. \right.$$

$$\times \left[e^{-((L_{\mathrm{in}}+w)+\vec{a}_k x_0 + \vec{b}_k k_0 t)^2/(1+t^2)} - e^{-((L_{\mathrm{in}}+w) - \vec{a}_k x_0 - \vec{b}_k k_0 t)^2/(1+t^2)} \right]$$

$$+ 8^{-1} \left(1 + 2b^2 k_0^2 \right) \left[\mathrm{erf}\left(\frac{(L_{\mathrm{in}} + w) + ax_0 + bk_0 t}{\sqrt{1+t^2}} \right) \right.$$

$$+ \mathrm{erf}\left(\frac{(L_{\mathrm{in}} + w) - ax_0 - bk_0 t}{\sqrt{1+t^2}} \right) \right]$$

$$\times \prod_{j=1, j\neq k}^{N} \left[\mathrm{erf}\left(\frac{(L_{\mathrm{in}} + w) + \vec{b}_j k_0 t + \vec{a}_j x_0}{\sigma\sqrt{1 + t^2/\sigma^4}} \right) \right.$$

$$\left. - \mathrm{erf}\left(\frac{-(L_{\mathrm{in}} + w) + \vec{b}_j k_0 t + \vec{a}_j x_0}{\sigma\sqrt{1 + t^2/\sigma^4}} \right) \right] \right\}$$

$$(6.1.7b)$$

These formulas were found by a Maple computation. Similar formulas not involving Herf^s can be found for $s \geq 2$ by maple as well, but there is no need to list them here.

6.2 Location of Each Framelet

We now introduce the bounding box, which we use to pinpoint the location of each framelet after it is propagated under the free flow, $e^{i(1/2)\Delta t}$. Intuitively, we are treating each framelet as a classical particle which has a finite radius which varies with time.

Definition 6.2.1 The collection of sets $\{\text{BB}_{(\vec{a},\vec{b},\sigma)}(\varepsilon, t)\}$ (indexed by $(\vec{a}, \vec{b}) \in \mathbb{Z}^N \times \mathbb{Z}^N$, $\varepsilon \in \mathbb{R}^+$, $t \in \mathbb{R}$) is a family of bounding boxes if:

$$\left\| e^{i(1/2)\Delta t} \phi_{(\vec{a},\vec{b})}(\vec{x}) \right\|_{H^s(\text{BB}_{(\vec{a},\vec{b},\sigma)}(\varepsilon,t)^C)} \leq \varepsilon \tag{6.2.1}$$

In particular, if $\text{BB}_{(\vec{a},\vec{b},\sigma)}(\varepsilon, t)$ is a collection of balls having radii which do not vary with $\vec{a}$, we let $\mathfrak{W}^s(\vec{b}, \varepsilon, t)$ denote the radius.

We also let $\mathfrak{w}_i^s(\vec{b}, \varepsilon)$, $\mathfrak{w}_v^s(\vec{b}, \varepsilon)$ denote the initial radius and the rate of dispersion, respectively, so that:

$$\mathfrak{W}^s(\vec{b}, \varepsilon, t) \leq \mathfrak{w}_i^s(\vec{b}, \varepsilon) + \mathfrak{w}_v^s(\vec{b}, \varepsilon)t \tag{6.2.2a}$$

$$\lim_{t \to \infty} \frac{\mathfrak{W}^s(\vec{b}, \varepsilon, t)}{\mathfrak{w}_v^s(\vec{b}, \varepsilon)t} = 1 \tag{6.2.2b}$$

Remark 6.2.2 We only prove that the numbers $\mathfrak{w}_i^s(\vec{b}, \varepsilon)$, $\mathfrak{w}_v^s(\vec{b}, \varepsilon)$ satisfying (6.2.2) exist for $s = 0, 1$ (c.f. Proposition 6.2.6). However, we believe it is intuitively clear that they will exist for any $s \in \mathbb{N}$, and that they could be found by doing calculations similar to those used in the proof of Proposition 6.2.6.

We now state a pair of Lemmas which demonstrate the usefulness of bounding boxes. This results show that to determine whether a given framelet is in BAD or NECC, it suffices to track it's bounding box. They are each formulated in somewhat technical terms. But the basic idea is this: if the distance between the classical center of mass of the framelet and the interior box is greater than the spreading of the framelet, the framelet is not relevant. Similarly, if the distance between the classical center of mass and the exterior of the computational boxx is less than the spreading of the framelet, the framelet is not bad.

Lemma 6.2.3 *Fix $T > 0$. Then the following two implications hold:*

(a) *Suppose, for $t \in [0, T]$, that $\mathrm{BB}_{(\vec{a},\vec{b},\sigma)}(\varepsilon, t) \cap [-L_{\mathrm{in}}, L_{\mathrm{in}}]^N = \emptyset$ (or $\mathrm{BB}_{(\vec{a},\vec{b},\sigma)}(\varepsilon, t) \cap [-L_{\mathrm{in}}, L_{\mathrm{in}}]^N$ has measure 0). Then $(\vec{a}, \vec{b}) \notin \mathrm{NECC}(\varepsilon, s, T)$.*

(b) *Suppose, for $t \in [0, T]$, that $\mathrm{BB}_{(\vec{a},\vec{b},\sigma)}(\varepsilon, t) \subset [-(L_{\mathrm{in}} + w), (L_{\mathrm{in}} + w)]^N$ (or $\mathrm{BB}_{(\vec{a},\vec{b},\sigma)}(\varepsilon, t) \cap ([-(L_{\mathrm{in}} + w), (L_{\mathrm{in}} + w)]^N)^C$ has measure 0). Then $(\vec{a}, \vec{b}) \notin \mathrm{BAD}(\varepsilon, s, T)$.*

Proof

(a) If $\mathrm{BB}_{(\vec{a},\vec{b},\sigma)}(\varepsilon, t) \cap [-L_{\mathrm{in}}, L_{\mathrm{in}}]^N = \emptyset$ (after possibly ignoring a set of measure 0), then $[-L_{\mathrm{in}}, L_{\mathrm{in}}]^N \subset \mathrm{BB}_{(\vec{a},\vec{b},\sigma)}(\varepsilon, t)^C$. Therefore:

$$\mathcal{R}^s_{(\vec{a},\vec{b})}(t) = \left\| e^{i(1/2)\Delta t} \phi_{(\vec{a},\vec{b})}(\vec{x}) \right\|_{H^s([-L_{\mathrm{in}}, L_{\mathrm{in}}]^N)}$$

$$\leq \left\| e^{i(1/2)\Delta t} \phi_{(\vec{a},\vec{b})}(\vec{x}) \right\|_{H^s(\mathrm{BB}_{(\vec{a},\vec{b},\sigma)}(\varepsilon,t)^C)} \leq \epsilon$$

where the last step follows by the definition of $\mathrm{BB}_{(\vec{a},\vec{b},\sigma)}(\varepsilon, t)$. Thus, $\mathcal{R}^s_{(\vec{a},\vec{b})}(t) \leq \epsilon$ for $t \in [0, T]$ and $(\vec{a}, \vec{b}) \notin \mathrm{NECC}(\varepsilon, s, T)$.

(b) If $\mathrm{BB}_{(\vec{a},\vec{b},\sigma)}(\varepsilon, t) \subset [-(L_{\mathrm{in}} + w), (L_{\mathrm{in}} + w)]^N$, then $([-(L_{\mathrm{in}} + w), (L_{\mathrm{in}} + w)]^N)^C \subset \mathrm{BB}_{(\vec{a},\vec{b},\sigma)}(\varepsilon, t)^C$. By Theorem 6.0.1, we find that:

$$\mathcal{E}^s_{(\vec{a},\vec{b})}(t) \leq \left\| e^{i(1/2)\Delta t} \phi_{(\vec{a},\vec{b})}(\vec{x}) \right\|_{H^s(([-(L_{\mathrm{in}}+w),(L_{\mathrm{in}}+w)]^N)^C)}$$

$$\leq \left\| e^{i(1/2)\Delta t} \phi_{(\vec{a},\vec{b})}(\vec{x}) \right\|_{H^s(\mathrm{BB}_{(\vec{a},\vec{b},\sigma)}(\varepsilon,t)^C)} \leq \epsilon$$

Again, the last step follows by the definition of $\mathrm{BB}_{(\vec{a},\vec{b},\sigma)}(\varepsilon, t)$. Thus $(\vec{a}, \vec{b}) \notin \mathrm{BAD}(\varepsilon, s, T)$.

$\square$

Lemma 6.2.4 *Fix $T > 0$. Then the following two implications hold.*

(a) *Suppose, for $t \in [0, T]$, that $d(\vec{a}x_0 + \vec{b}k_0 t, [-L_{\mathrm{in}}, L_{\mathrm{in}}]^N) \geq \mathfrak{w}^s_i(\vec{b}, \epsilon) + \mathfrak{w}^s_v(\vec{b}, \epsilon)t$. Then $(\vec{a}, \vec{b}) \notin \mathrm{NECC}(\varepsilon, s, T)$.*

(b) *Suppose, for $t \in [0, T]$, that $|\vec{a}x_0 + \vec{b}k_0 t|_\infty \leq \mathfrak{w}^s_i(\vec{b}, \epsilon) + \mathfrak{w}^s_v(\vec{b}, \epsilon)t$. Then $(\vec{a}, \vec{b}) \notin \mathrm{BAD}(\varepsilon, s, T)$.*

Proof

(a) If $d(\vec{a}x_0 + \vec{b}k_0 t, [-L_{\mathrm{in}}, L_{\mathrm{in}}]^N) \geq \mathfrak{w}^s_i(\vec{b}, \epsilon) + \mathfrak{w}^s_v(\vec{b}, \epsilon)t$, then

$$\mathrm{interior}\{\mathrm{BB}_{(\vec{a},\vec{b},\sigma)}(\varepsilon, t)\} \cap [-L_{\mathrm{in}}, L_{\mathrm{in}}]^N = \emptyset$$

Since the boundary of $\mathrm{BB}_{(\vec{a},\vec{b},\sigma)}(\varepsilon, t)$ has measure 0, we find that $(\vec{a}, \vec{b})$ satisfies Lemma 6.2.3, part (a).

(b) The same idea applis, except now:

$$\text{interior}\{\mathrm{BB}_{(\vec{a},\vec{b},\sigma)}(\varepsilon, t)\} \cap ([-(L_{\text{in}} + w), (L_{\text{in}} + w)]^N)^C = \emptyset$$

This, combined with Lemma 6.2.3, part (b) yields the result we seek.

□

We now calculate precisely a bounding box in the spaces L^2 and H^1. We recall first the complementary incomplete Gamma function, and define it's partial inverse.

Definition 6.2.5 The complementary incomplete Gamma function, $\Gamma(a, x)$ is defined by:

$$\Gamma(a, x) = \int_x^\infty e^{-t} t^{a-1} dt \tag{6.2.3}$$

It has the asymptotic behavior:

$$\Gamma(a, x) \sim x^{a-1} e^{-x} \sum_{j=0}^\infty \frac{(a-1)(a-2)\dots(a-j)}{x^j} \tag{6.2.4}$$

Moreover, if $n \geq a - 1$, we find that:

$$\left| \Gamma(a, x) - x^{a-1} e^{-x} \sum_{j=0}^n \frac{(a-1)(a-2)\dots(a-j)}{x^j} \right|$$

$$\leq x^{a-1} e^{-x} \frac{(a-1)(a-2)\dots(a-(n+1))}{x^{n+1}} \tag{6.2.5}$$

We define the partial inverse of the complementary incomplete Gamma function, $\Gamma^{-1}(a, x)$ to be the inverse of the function $\mathbb{R}^+ \ni x \mapsto \Gamma(a, x)$ for fixed a, so that $\Gamma(a, \Gamma^{-1}(a, x)) = x$.

Note that because $\Gamma(a, x)$ is monotone decreasing in x for a real, $\Gamma^{-1}(a, \epsilon)$ is monotonically increasing as $\epsilon \to 0$. The rate of increase is slower than ϵ^{-t} for any $t > 0$.

Proposition 6.2.6 *The following family forms a collection of bounding boxes:*

$$\mathrm{BB}_{(\vec{a},\vec{b},\sigma)}(\varepsilon, t) = B_{\mathfrak{W}^s(\vec{b},\epsilon,t)}(\vec{a} x_0 + \vec{b} k_0 t) \tag{6.2.6}$$

For s = 0, 1, $\mathfrak{W}^s(\vec{b}, \epsilon, t)$ is given by:

$$\mathfrak{W}^0(\vec{b}, \epsilon, t) = \sqrt{\sigma^2 + t^2/\sigma^2}(\Gamma^{-1}(N/2, 2\epsilon^2\pi^{N/2}/\left|S^{N-1}\right|))^{1/2} \qquad (6.2.7a)$$

$$\mathfrak{W}^1(\vec{b}, \epsilon, t) = \sqrt{\sigma^2 + t^2/\sigma^2} \max\left\{\left[\Gamma^{-1}\left(N/2, \frac{\epsilon^2\pi^N/2}{2\left|S^{N-1}\right|(1 + |\vec{b}k_0|_2^2)}\right)\right]^{1/2},\right.$$

$$\left.\left[\Gamma^{-1}\left((N+2)/2, \frac{\epsilon^2\sigma^2\pi^{N/2}}{2\left|S^{N-1}\right|}\right)\right]^{1/2}\right\} \qquad (6.2.7b)$$

Here, $\left|S^{N-1}\right|$ is the angular measure of the unit ball. We also find that:

$$\mathfrak{w}_i^0(\vec{b}, \epsilon) = \sigma\sqrt{\Gamma^{-1}(N/2, 2\epsilon^2\pi^{N/2}/\left|S^{N-1}\right|)}$$

$$\mathfrak{w}_v^0(\vec{b}, \epsilon) = \sigma^{-1}(\Gamma^{-1}(N/2, 2\epsilon^2\pi^{N/2}/\left|S^{N-1}\right|))^{1/2} \qquad (6.2.8a)$$

$$\mathfrak{w}_i^1(\vec{b}, \epsilon) = \sigma \max\left\{\left[\Gamma^{-1}\left(N/2, \frac{\epsilon^2\pi^N/2}{2\left|S^{N-1}\right|(1 + |\vec{b}k_0|_2^2)}\right)\right]^{1/2},\right.$$

$$\left.\left[\Gamma^{-1}\left((N+2)/2, \frac{\epsilon^2\sigma^2\pi^{N/2}}{2\left|S^{N-1}\right|}\right)\right]^{1/2}\right\}$$

$$\mathfrak{w}_v^1(\vec{b}, \epsilon) = \sigma^{-1} \max\left\{\left[\Gamma^{-1}\left(N/2, \frac{\epsilon^2\pi^N/2}{2\left|S^{N-1}\right|(1 + |\vec{b}k_0|_2^2)}\right)\right]^{1/2},\right.$$

$$\left.\left[\Gamma^{-1}\left((N+2)/2, \frac{\epsilon^2\sigma^2\pi^{N/2}}{2\left|S^{N-1}\right|}\right)\right]^{1/2}\right\}$$

Proof A straightforward computation, similar to the previous results. The main difference is that we work in spherical, rather than rectangular coordinates. To begin, change variables to $\vec{z}(t) = (\vec{x} - \vec{b}k_0t - \vec{a}x_0)/\sqrt{\sigma^2 + t^2/\sigma^2}$. We note that $d\vec{x} = (\sigma^2 + t^2/\sigma^2)^{N/2}d\vec{z}$. In this new coordinate system, we find that:

$$\left|e^{i(1/2)\Delta t}\phi_{(\vec{a},\vec{b})}(\vec{x})\right|^2 = \pi^{-N/2}(\sigma^2 + t^2/\sigma^2)^{-N/2}e^{-z^2} \qquad (6.2.9)$$

Thus:

$$\int \left| e^{i(1/2)\Delta t} \phi_{(\vec{a},\vec{b})}(\vec{x}) \right|^2 d\vec{x}$$

$$= \int \left| \frac{\exp(i\vec{b}k_0 \cdot (\vec{x} - \vec{b}k_0/2t - \vec{a}x_0))}{\pi^{N/4}\sigma^{N/2}(1 + it/\sigma^2)^{N/2}} \exp\left(\frac{-|\vec{x} - \vec{b}k_0 t - \vec{a}x_0|_2^2}{2\sigma^2(1 + it/\sigma^2)} \right) \right|^2 d\vec{x}$$

$$= \int \pi^{-N/2} e^{-|\vec{z}|_2^2} d\vec{z} \qquad (6.2.10)$$

Note that the domain of integration also needs to be altered, but we suppressed this to simplify (6.2.10).

Bounding Boxes in L^2

We switch to polar coordinates about the center of mass of the framelet, $\vec{z} = r\vec{\Omega}$, where $\Omega \in S^{N-1}$.

Thus, if we integrate outside a ball of radius R around the point $\vec{z}(t)$, we obtain:

$$\int_{S^{N-1}} \int_R^\infty \pi^{-N/2} e^{-r^2} r^{N-1} dr d\vec{\Omega} = \pi^{-N/2} \left| S^{N-1} \right| \int_{R^2}^\infty u^{(N-1)/2} e^{-u} \frac{du}{2\sqrt{u}}$$

$$= (1/2)\pi^{-N/2} \left| S^{N-1} \right| \int_{R^2}^\infty u^{N/2-1} e^{-u} du$$

$$= (1/2)\pi^{-N/2} \left| S^{N-1} \right| \Gamma(N/2, R^2) \qquad (6.2.11)$$

where $\Gamma(a, x)$ is the incomplete Gamma function (c.f. [1]).

Therefore, if $R = (\Gamma^{-1}(N/2, 2\epsilon^2 \pi^{N/2}/|S^{N-1}|))^{1/2}$, then we find that (6.2.11) $\leq \epsilon^2$. Backtracking, this implies that in the $\vec{z}(t)$ coordinate system, the bounding box is a ball of radius $(\Gamma^{-1}(N/2, 2\epsilon^2 \pi^{N/2}/|S^{N-1}|))^{1/2}$. In the $\vec{x}$ coordinate system, this implies that the bounding box is a ball of radius $\sqrt{\sigma^2 + t^2/\sigma^2}(\Gamma^{-1}(N/2, 2\epsilon^2 \pi^{N/2}/|S^{N-1}|))^{1/2}$ around the point $\vec{a}x_0 + \vec{b}k_0 t$.

Bounding Boxes in H^1

The main difference between L^2 and H^1 is that in H^1, we need to compute:

$$\int \left| e^{i(1/2)\Delta t} \phi_{(\vec{a},\vec{b})}(\vec{x}) \right|^2 + \sum_{j=1}^N \left| \partial_{x_j} e^{i(1/2)\Delta t} \phi_{(\vec{a},\vec{b})}(\vec{x}) \right|^2 d\vec{x}$$

We begin by computing $\partial_{x_j} e^{i(1/2)\Delta t} \phi_{(\vec{a},\vec{b})}(\vec{x})$.

$$\partial_{x_j} \frac{\exp(i\vec{b}k_0 \cdot (\vec{x} - \vec{b}k_0/2t - \vec{a}x_0))}{\pi^{N/4}\sigma^{N/2}(1 + it/\sigma^2)^{N/2}} \exp\left(\frac{-|\vec{x} - \vec{b}k_0 t - \vec{a}x_0|_2^2}{2\sigma^2(1 + it/\sigma^2)}\right)$$

$$= \left(i\vec{b}_j k_0 - \frac{|\vec{x}_j - \vec{b}_j k_0 t - \vec{a}_j x_0|}{\sigma^2(1 + it/\sigma^2)}\right)$$

$$\times \frac{\exp(i\vec{b}k_0 \cdot (\vec{x} - \vec{b}k_0/2t - \vec{a}x_0))}{\pi^{N/4}\sigma^{N/2}(1 + it/\sigma^2)^{N/2}} \exp\left(\frac{-|\vec{x} - \vec{b}k_0 t - \vec{a}x_0|_2^2}{2\sigma^2(1 + it/\sigma^2)}\right) \qquad (6.2.12)$$

We take the absolute square of this, to obtain:

$$\left(|\vec{b}_j k_0|^2 + \frac{|\vec{x}_j - \vec{b}_j k_0 t - \vec{a}_j x_0|^2}{\sigma^4(1 + t^2/\sigma^4)} + \text{antisymmetric terms}\right)$$

$$\times \pi^{-N/2}(\sigma^2 + t^2/\sigma^2)^{-N} \exp\left(\frac{|\vec{x} - \vec{b}k_0 t - \vec{a}x_0|_2^2}{\sigma^2(1 + t^2/\sigma^4)}\right) \qquad (6.2.13)$$

The antisymmetric terms are antisymmetric about the point $\vec{x} = \vec{a}_j x_0 + \vec{b}_j k_0 t$. Thus, upon integration in $\vec{x}$, these terms will vanish.

We then add this up for $j = 1 \ldots N$, and add a constant term. This implies that:

$$\pi^{-N/2}(\sigma^2 + t^2/\sigma^2)^{-N} \exp\left(\frac{|\vec{x} - \vec{b}k_0 t - \vec{a}x_0|_2^2}{\sigma^2(1 + t^2/\sigma^4)}\right)$$

$$+ \sum_{j=1}^{N}\left(|\vec{b}_j k_0|^2 + \frac{|\vec{x}_j - \vec{b}_j k_0 t - \vec{a}_j x_0|^2}{\sigma^4(1 + t^2/\sigma^4)} + \text{antisymmetric terms}\right)$$

$$\times \pi^{-N/2}(\sigma^2 + t^2/\sigma^2)^{-N} \exp\left(\frac{|\vec{x} - \vec{b}k_0 t - \vec{a}x_0|_2^2}{\sigma^2(1 + t^2/\sigma^4)}\right)$$

$$= \pi^{-N/2}(\sigma^2 + t^2/\sigma^2)^{-N}(1 + |\vec{b}k_0|_2^2) \exp\left(\frac{|\vec{x} - \vec{b}k_0 t - \vec{a}x_0|_2^2}{\sigma^2(1 + t^2/\sigma^4)}\right)$$

$$+ \frac{|\vec{x} - \vec{b}k_0 t - \vec{a}x_0|_2^2}{\sigma^4(1 + t^2/\sigma^4)} \exp\left(\frac{|\vec{x} - \vec{b}k_0 t - \vec{a}x_0|_2^2}{\sigma^2(1 + t^2/\sigma^4)}\right) \qquad (6.2.14)$$

Switching to the $\vec{z}$ coordinate system yields:

$$(6.2.14) = \pi^{-N/2}(\sigma^2 + t^2/\sigma^2)^{-N}(1 + |\vec{b}k_0|_2^2)e^{-|\vec{z}|_2^2}$$

$$+ \pi^{-N/2}(\sigma^2 + t^2/\sigma^2)^{-N}\sigma^{-2}|\vec{z}|_2^2 e^{-|\vec{z}|_2^2} \qquad (6.2.15)$$

Switching again to polar coordinates and integrating out the angular part yields:

$$\int_R^\infty (6.2.15)d\vec{z}$$

$$= \pi^{-N/2}\left|S^{N-1}\right|\left((1 + |\vec{b}k_0|_2^2)\int_R^\infty e^{-r^2}r^{N-1}dr + \sigma^{-2}\int_R^\infty r^2 e^{-r^2}r^{N-1}dr\right)$$

$$= (1/2)\pi^{-N/2}\left|S^{N-1}\right|\left((1 + |\vec{b}k_0|_2^2)\Gamma(N/2, R^2) + \sigma^{-2}\Gamma(N/2+1, R^2)\right)$$

$$\qquad (6.2.16)$$

If R^2 satisfies

$$R^2 \geq \max\left\{\Gamma^{-1}\left(N/2, \frac{\epsilon^2\pi^N/2}{2\left|S^{N-1}\right|(1 + |\vec{b}k_0|_2^2)}\right),\right.$$

$$\left.\Gamma^{-1}\left((N+2)/2, \frac{\epsilon^2\sigma^2\pi^{N/2}}{2\left|S^{N-1}\right|}\right)\right\} \qquad (6.2.17)$$

then:

$$(1/2)\pi^{-N/2}\left|S^{N-1}\right|(1 + |\vec{b}k_0|_2^2)\Gamma(N/2, R^2) \leq \epsilon^2/2$$

$$(1/2)\pi^{-N/2}\left|S^{N-1}\right|\Gamma(N/2+1, R^2)\sigma^{-2} \leq \epsilon^2/2$$

and thus $(6.2.16) \leq \epsilon^2$.

Switching from polar coordinates to $\vec{z}$ coordinates implies that the bounding box consists of $B_R(0)^C$ with R satisfying (6.2.17). Changing coordinates once more to $\vec{x}$ yields the result we seek.

To obtain (6.2.8), we simply observe that $\sqrt{\sigma^2 + t^2/\sigma^2} \leq \sigma + t/\sigma$ for $t > 0$, and apply this to (6.2.7).

$\square$

Similar computations can be done in H^s for $s > 1$, but we neglect to do them here.

We now state one more result, which we will use later.

Proposition 6.2.7 *Let* $s = 0, 1$. *Then:*

$$\sup_{\vec{b} \in \mathrm{LF}(K)} \mathfrak{W}^{\epsilon}(\vec{b}, 0, t) = \sqrt{\sigma^2 + t^2/\sigma^2} (\Gamma^{-1}(N/2, 2\epsilon^2 \pi^{N/2} / |S^{N-1}|))^{1/2}$$

$$(6.2.18a)$$

$$\sup_{\vec{b} \in \mathrm{LF}(K)} \mathfrak{W}^{\epsilon}(\vec{b}, 1, t) = \sqrt{\sigma^2 + t^2/\sigma^2}$$

$$\times \max \left\{ \left[\Gamma^{-1}\left(N/2, \frac{\epsilon^2 \pi^N/2}{2 |S^{N-1}| (1 + NK^2)}\right) \right]^{1/2}, \right.$$

$$\left. \left[\Gamma^{-1}\left((N+2)/2, \frac{\epsilon^2 \sigma^2 \pi^{N/2}}{2 |S^{N-1}|}\right) \right]^{1/2} \right\} \qquad (6.2.18b)$$

Note the result is independent of K for $s = 0$.

Proof If $s = 0$, $\mathfrak{W}^{\epsilon}(\vec{b}, 0, t)$ does not vary with $\vec{b}$. This proves (6.2.18a).

We now prove (6.2.18b). This follows simply because if $t < t'$, $\Gamma^{-1}(a, t) \geq \Gamma^{-1}(a, t')$ for any $a \in \mathbb{R}^+$. Applying this to (6.2.7b), we find that the sup on the left side of (6.2.18b) is maximized when $|\vec{b}k_0|_2$ is maximized. This occurs when $\vec{b}_j k_0 = \lfloor K \rfloor$. Thus, $|\vec{b}k_0|_2 \leq \sqrt{N}K$, and we obtain the bound we seek.   □

Chapter 7
Assumptions and Accuracy Estimates

In this section, we prove the accuracy of our method, subject to some assumptions on the equation.

We do not prove a complete error bound. Let $\Psi(\vec{x}, t)$ be the solution to (1.1.1) on $\mathbb{R}^N$ and let $\Phi(\vec{x}, t)$ be the approximate solution generated by our algorithm (defined on $[-L_{\text{trunc}}, L_{\text{trunc}}]^N$).

To obtain complete control on the error (letting $\Phi_d(\vec{x}, t)$ be the discretized version of $\Phi(\vec{x}, t)$), we need to control:

$$\|\Psi(\vec{x}, t) - \Phi_d(\vec{x}, t)\|_{H_b^s} \leq \|\Psi(\vec{x}, t) - \Phi_d(\vec{x}, t)\|_{H_b^s} + \|\Phi(\vec{x}, t) - \Phi_d(\vec{x}, t)\|_{H_b^s}$$

We only prove a bound on the first term, $\|\Psi(\vec{x}, t) - \Psi_b(\vec{x}, t)\|_{H_b^s}$.

Bounds on the second term depend crucially on many details of the implementation. That is, they will vary depending on whether one chooses a finite element, finite difference or spectral method. They will vary with the timestep, space discretization and also floating point (or other roundoff) error. We assume this is known and is also sufficiently small as to be negligible.

Our goal, is to reduce the error caused by $\|\Psi(\vec{x}, t) - \Phi(\vec{x}, t)\|_{H_b^s}$ to the same order of magnitude as the discretization error, $\|\Phi(\vec{x}, t) - \Phi_d(\vec{x}, t)\|_{H_b^s}$.

7.1 Assumptions

Let us assume we wish to solve (1.1.1) on a time interval $[0, T_{\mathfrak{M}}]$ with error ε measured in a Sobolev space $H^s([-L_{\text{in}}, L_{\text{in}}]^N)$. We now state our assumptions.

Assumption 1 *We assume the solution to (1.1.1) exists and is unique on $\mathbb{R}^N$ for $t \in [0, T_{\mathfrak{M}}]$. We denote by $\mathfrak{U}(t)$ the propagator on $\mathbb{R}^N$.*

© The Author(s), under exclusive license to Springer Nature Singapore Pte Ltd. 2023
A. Soffer et al., *Time Dependent Phase Space Filters*, SpringerBriefs on PDEs and Data Science, https://doi.org/10.1007/978-981-19-6818-1_7

In particular, we assume that there exists a function $\mathcal{L}(t)$ and a large number M such that for all $\Psi_0(x)$ with $\|\Psi_0(x)\|_{H^s} \leq M$:

$$\|\mathfrak{U}(t)\Psi_0(x) - \mathfrak{U}(t)\Psi_1(x)\|_{\mathcal{L}(H^s, H^s)} \leq \mathcal{L}(t)\,\|\Psi_0(x) - \Psi_1(x)\|_{H^s} \qquad (7.1.1)$$

Assumption 2 *There exists a maximal momentum $k_{\sup} = k_{\sup}(\Psi_0)$ in the following sense. For all $t \in [0, T_{\mathfrak{M}}]$, $\delta_{\sup} > 0$, we can compute a $k_{\sup}(\delta_{\sup})$ such that:*

$$\sup_{t \in [0, T_{\mathfrak{M}}]} \left\|\mathcal{P}_{HF(k_{\sup})}\Psi(\vec{x}, t)\right\| < \delta_{\sup} \qquad (7.1.2)$$

Assumption 3 *The nonlinearity is Lipschitz in H^s. That is, there exists a constant $\mathcal{G}$ such that for $u, v \in H^s$:*

$$\|g(t, \vec{x}, u)u - g(t, \vec{x}, v)v\|_{H^s} \leq \mathcal{G}\,\|u - v\|_{H^s} \qquad (7.1.3)$$

Although many common nonlinearities are not Lipschitz, they are typically locally Lipschitz, c.f. [25, section 3.2]. Therefore most nonlinearities of interest can be modified appropriately to satisfy these assumptions.

Assumption 4 *The nonlinearity $g(t, \vec{x}, \Psi)\Psi$ is well localized in phase space. That is, for any $\delta_{NL} > 0$, there exist constants $L_{NL} = L_{NL}(\delta_{NL})$ and $k_{\sup,NL} = k_{\sup,NL}(\delta_{NL})$ (uniform on $t \in [0, T]$) such that:*

$$\left\|\mathcal{P}_{NL^c}g(t, \vec{x}, \Psi(\vec{x}, t))\Psi(\vec{x}, t)\right\|_{H^s} < \delta_{NL}\,\|\Psi(\vec{x}, t)\|_{H^s} \qquad (7.1.4a)$$

$$NL = \{(\vec{a}, \vec{b}) \in \mathbb{Z}^N \times \mathbb{Z}^N : |\vec{a}|_\infty \leq L_{NL}(\delta_{NL}) \text{ and } \left|\vec{b}\right|_\infty \leq k_{\sup,NL}(\delta_{NL})\} \qquad (7.1.4b)$$

Assumption 5 *We assume that for each $\delta_F > 0$, we can find an ϵ, a function $\mathfrak{W}^+(\vec{b}, \delta_F, t) \leq \mathfrak{W}^s(\vec{b}, \epsilon, t)$ and $L_F = L_F(\delta_F)$ so that if*

$$d(\vec{a}x_0 + \vec{b}k_0 t, [-L_F, L_F]^N) > \mathfrak{W}^+(\vec{b}, \delta_F, t)$$

for all $(\vec{a}, \vec{b}) \in F \subseteq \mathbb{Z}^N \times \mathbb{Z}^N$, then $\mathcal{P}_F\Psi$ propagates essentially freely. By this we mean that:

$$\sup_{t \in [0, T_{\mathfrak{M}}]} \left\|\mathfrak{U}(t|\Psi_0)\mathcal{P}_F\Psi(x) - e^{i(1/2)\Delta t}\mathcal{P}_F\Psi(x)\right\|_{H_b^s} \leq \delta_F\,\|\mathcal{P}_F\Psi(x)\|_{H_b^s} \qquad (7.1.5a)$$

$$\sup_{t \in [0, T_{\mathfrak{M}}]} \left\|\mathfrak{U}(t|\Psi_0)\mathcal{P}_{F^c}\Psi_0(x) - \mathfrak{U}(t|\mathcal{P}_{F^c}\Psi_0)\mathcal{P}_{F^c}\Psi_0(x)\right\|_{H_b^s}$$
$$\leq \mathcal{L}_{ext}(t)\,\|\mathcal{P}_F\Psi(x)\|_{H_b^s} \qquad (7.1.5b)$$

The function $\mathcal{L}_{ext}(t)$ must satisfy $\mathcal{L}_{ext}(0) = 0$ and $\sup_{t \in [0, T_{\mathfrak{M}}]} \mathcal{L}_{ext}(t) = \delta_F$.

Remark 7.1.1 Assumption 5 says that outside of a certain box in phase space, the problem is approximately linear the free propagator is sufficiently accurate. This assumption will be the most difficult assumption to verify in the nonlinear case, though it is verified for some linear potentials in Sect. 8.6.

The condition that $\mathfrak{W}^+(\vec{b}, \delta_F, t) \leq \mathfrak{W}^s(\vec{b}, \epsilon, t)$ is present to to ensure that $\mathfrak{W}^+(\vec{b}, \delta_F, t)$ is non-trivial, that is to ensure that $\mathfrak{W}^+(\vec{b}, \delta_F, t)$ is directly related to how much mass enters the region $[-L_F, L_F]^N$.

Remark 7.1.2 In principle we can use $\mathcal{L}(t)$ as a bound on $\mathcal{L}_{\text{ext}}(t)$, but this is far from optimal. $\mathcal{L}_{\text{ext}}(t)$ should be small for relatively long times, while $\mathcal{L}(t)$ may not be. In the linear case, $\mathcal{L}_{\text{ext}}(t) = 0$.

Assumption 6 *We assume that mass does not pile up on tangential, slow waves or returning waves in the following sense. We assume there exists a $k_{\text{inf}} = k_{\text{inf}}(\delta_{\text{inf}})$, $L_{\text{inf}} = L_{\text{inf}}(\delta_{\text{inf}})$ such that:*

$$\sup_{t \in [0, T_{\mathfrak{M}}]} \|\mathcal{P}_S \Psi(\vec{x}, t)\|_{H_b^s} < \delta_{\text{inf}} \tag{7.1.6}$$

with S a set satisfying:

$$\forall (\vec{a}, \vec{b}) \in S, !(\exists j, |\vec{a}_j x_0| \geq L_{\text{inf}} \text{ and } \vec{b}_j k_0(\vec{a}_j / |\vec{a}_j|) > k_{\text{inf}})$$

$$\text{and } |\vec{a} x_0|_\infty \geq L_{\text{inf}} \tag{7.1.7}$$

Essentially, what Assumption 6 is saying is that most of the waves outside the box $[-L_{\text{inf}}, L_{\text{inf}}]^N$ are moving faster than some small velocity k_{inf}, and are moving outward (away from $[-L_{\text{inf}}, L_{\text{inf}}]^N$).

7.2   Discussions on the Assumptions

It is simple to observe that Gronwall's lemma combined with Assumption 3 implies Assumption 1 with $\mathcal{L}(t) = \mathcal{G}e^{\mathcal{G}t}$. However, if better estimates are available, they should be used, since the error estimates we give will be given in terms of $\mathcal{L}(t)$.

Although we state our assumptions in terms of WFT coefficients, they are actually just rephrased versions of more standard assumptions. We provide here some sufficient conditions for verifying the more technical assumptions.

Proposition 7.2.1 (Sufficient Conditions for Assumption 2) *Suppose that there exists a maximal momentum $k_{\text{sup}} = k_{\text{sup}}(\Psi_0)$ in the following sense. For all $t \in [0, T_{\mathfrak{M}}]$, $\delta_{\text{sup}} > 0$, we can compute a $k_{\text{sup}}(\delta_{\text{sup}})$ such that:*

$$\left\| P^0_{[-K', K']^N; 0}(k)\Psi(\vec{x}, t) \right\|_{H^s} < \delta_{\text{sup}} / (2\mathfrak{H}^s_+(\tilde{g}(\vec{x}))\mathfrak{H}^{-s}_+(e^{-x^2/\sigma^2})) \tag{7.2.1}$$

with $K' = k_{\text{sup}} - \mathfrak{K}^s(K)$. Then Assumption 2 holds.

Proof Merely apply Theorem 3.3.3.   □

Proposition 7.2.2 (Sufficient Conditions for Assumption 4) *Suppose that tThe nonlinearity $g(t, \vec{x}, \Psi)\Psi$ is well localized in phase space in the traditional sense. That is, for any $\delta_{NL} > 0$, there exist constants $L'_{NL} = L'_{NL}(\delta_{NL})$ and $k'_{\text{sup,NL}} = k'_{\text{sup,NL}}(\delta_{NL})$ (uniform on $t \in [0, T]$) such that:*

$$\left\| P^s_{[-k'_{\text{sup,NL}}, k'_{\text{sup,NL}}]^N; k_0}(\vec{x}) g(t, \vec{x}, \Psi(\vec{x}, t)) \Psi(\vec{x}, t) \right\|_{H^s}$$
$$< \frac{\delta_{NL} \|\Psi(\vec{x}, t)\|_{H^s}}{(4\mathfrak{H}^s_+(\tilde{g}(\vec{x}))\mathfrak{H}^{-s}_+(e^{-x^2/\sigma^2}))} \qquad (7.2.2a)$$

$$\left\| P^s_{[-L'_{NL}, L'_{NL}]^N; x_0}(\vec{x}) g(t, \vec{x}, \Psi(\vec{x}, t)) \Psi(\vec{x}, t) \right\|_{H^s}$$
$$< \frac{\delta_{NL} \|\Psi(\vec{x}, t)\|_{H^s}}{(4\mathfrak{H}^s_+(\tilde{g}(\vec{x}))\mathfrak{H}^{-s}_+(e^{-x^2/\sigma^2}))} \qquad (7.2.2b)$$

The constants L'_{NL} and $k'_{\text{sup,NL}}$ are related to those in Assumption 4 by the relations

$$L'_{NL} = L_{NL} - \mathfrak{X}^s_\square(\delta_{NL}/4\mathfrak{H}^s_+(\tilde{g}(\vec{x}))\mathfrak{H}^{-s}_+(e^{-x^2/\sigma^2}), k_{\text{sup,NL}}, L_{NL}) \quad (7.2.3)$$

$$k'_{\text{sup,NL}} = k_{\text{sup,NL}} - \mathfrak{K}^s_\square(\delta_{NL}/4\mathfrak{H}^s_+(\tilde{g}(\vec{x}))\mathfrak{H}^{-s}_+(e^{-x^2/\sigma^2}), k_{\text{sup,NL}}) \qquad (7.2.4)$$

Then Assumption 4 holds.

Proof Merely apply Theorem 3.3.7.   □

Assumption 5 states that framelets which are outgoing under the free flow are also outgoing under the full flow, which is very similar to a typical proof of scattering. There are two major differences, however. First, Assumption 5 is a finite time estimate, which makes proofs significantly easier since one takes a limit as $t \to T_{\mathfrak{M}}$ instead of $t \to \infty$. Second, Assumption 5 requires sharper control, in particular requiring quantitative control on L_F and $\mathfrak{W}^+(\vec{b}, \delta_F, t)$.

Assumption 6 is two statements. First, it assumes that the mass of the solution below some velocity k_{inf} is small. Second, it assumes that the solution stays on the "propagation set", that is the solution remains restricted to trajectories where $\vec{x} \parallel \vec{k}$. This assumption is really just a rephrasing of standard propagation estimates into the language of framelets.

Results along these lines are not uncommon in scattering theory, e.g. [39, 61], and are a common ingredient in proving dispersive estimates and asymptotic completness. However, we know of no such results which are proved in terms of framelets of any sort; rather, most are proved by constructing propagation observables. Furthermore, we are aware of no result with the quantitative control required by Assumption 6.

Most results along these lines do not control the generation of low frequencies, at least not without some assumption on the existence of a spectral gap (e.g. nonexistence of low energy bound states). Such assumptions are unsuitable for our purposes, since if we knew the spectrum of the Hamiltonian it would be pointless to run a numerical simulation! Nevertheless, we do expect this result to be true very often, and the algorithm is capable of detecting when it has become invalid. Thus, if the algorithm fails for this reason, the user is at least notified of the failure. Furthermore, this problem is resolved in [63], and the algorithm given there can be used if the TDPSF algorithm fails.

A stronger (but perhaps clearer) assumption than Assumption 6 would be the following:

Proposition 7.2.3 (Sufficient Condition for Assumption 6) *Let* $PS(L, k_{\inf})$ *(the propagation set) be defined by:*

$$PS(L, k_{\inf}) = \{(\vec{a}, \vec{b}) \in \mathbb{Z}^N \times \mathbb{Z}^N : |\vec{b}k_0|_2 > 2\sqrt{N}k_{\inf},$$
$$|\vec{b}k_0 - (|\vec{a}|_2^{-1}\vec{a}) \cdot \vec{b}k_0|_2 \le |\vec{b}k_0|_2/(4\sqrt{N})\}$$

Suppose that for any $\delta_{\inf}$, $\exists k_{\inf}$, $L_{\inf}$ *so that if* S *is a set satisfying*

$$S \cap PS(L_{\inf}, k_{\inf}) = \emptyset \tag{7.2.5a}$$
$$S \subseteq \{(\vec{a}, \vec{b}) \in \mathbb{Z}^N \times \mathbb{Z}^N : |\vec{a}x_0|_2 \ge L_{\inf}\}, \tag{7.2.5b}$$

then:

$$\sup_{t \in [0, T_{\mathfrak{M}}]} \|\mathcal{P}_S \Psi(\vec{x}, t)\|_{H_b^s} < \delta_{\inf} \tag{7.2.6}$$

Then Assumption 6 holds.

Proof We must show that any set S satisfying (7.1.7) also satisfies (7.2.5). This will show that the conditions of Proposition 7.2.3 imply Assumption 6.

Toward that end, let S be such a set. Since for any $(\vec{a}, \vec{b}) \in S$, $|\vec{a}x_0|_\infty \ge L_{\inf}$, we find that (7.2.5b) is satisfied. We must now show that:

$$\forall(\vec{a}, \vec{b}) \in S, !(\exists j, |\vec{a}_j x_0| \ge L_{\inf} \text{ and } \vec{b}_j k_0(\vec{a}_j/|\vec{a}_j|) > k_{\inf})$$

Equivalently:

$$\forall(\vec{a}, \vec{b}) \in S, \forall j, |\vec{a}_j x_0| \le L_{\inf} \text{ or } \vec{b}_j k_0(\vec{a}_j/|\vec{a}_j|) \le k_{\inf}$$

Now fix $(\vec{a}, \vec{b}) \in S$. We must show that $(\vec{a}, \vec{b}) \notin PS(L_{\inf}, k_{\inf})$. We proceed by contradiction.

Suppose $(\vec{a}, \vec{b}) \in \text{PS}(L_{\inf}, k_{\inf}) \cap S$. Define $\vec{z} = |\vec{a}|_2^{-1} \vec{a} \cdot \vec{b}k_0$. Then:

$$|\vec{z}|_2 \geq |\vec{b}k_0|_2 - |\vec{b}k_0 - \vec{z}|_2 \geq |\vec{b}k_0|_2 - |\vec{b}k_0|_2/(4\sqrt{N}) \tag{7.2.7}$$

Since $\vec{z}$ is a vector in the direction of $\vec{a}$, we find that $\exists j \in 1 \ldots N$ so that $|\vec{z}_j| \geq |\vec{z}|_2/\sqrt{N}$, and in addition, for this same j, $|\vec{a}_j x_0| \geq |\vec{a}x_0|_2/\sqrt{N}$. If j is chosen to be the component for which $|\vec{a}_j|$ is largest, then in addition $|\vec{a}_j x_0| \geq L_{\inf}$.

This implies that:

$$|\vec{z}_j| \geq |\vec{b}k_0|_2/\sqrt{N} - |\vec{b}k_0|_2/(4\sqrt{N})$$
$$|\vec{a}_j x_0| \geq L_{\inf}$$

In addition, the signs of $\vec{a}_j$ and $\vec{z}_j$ are the same. Now, observe that:

$$|\vec{b}_j k_0 - \vec{z}_j| \leq |\vec{b}k_0 - \vec{z}|_2 \leq |\vec{b}k_0|_2/(4\sqrt{N})$$

so:

$$|\vec{b}_j k_0| \geq |\vec{z}_j| - |\vec{b}k_0|_2/(4\sqrt{N}) \geq |\vec{b}k_0|_2/\sqrt{N} - 2[|\vec{b}k_0|_2/(4\sqrt{N})]$$
$$\geq |\vec{b}k_0|_2/(2\sqrt{N}) > k_{\inf}$$

But this contradicts (7.1.7), and also (7.2.7). Therefore there does not exist $(\vec{a}, \vec{b}) \in \text{PS}(L_{\inf}, k_{\inf}) \cap S$, and we are done. □

Remark 7.2.4 Without using framelets, statements such as those assumed in Proposition 7.2.3 are common if we ignore low frequencies. The following estimate holds when $g(t, \vec{x}, \Psi(x, t))\Psi(x, t) = 0$ (in which case $C = \|\vec{x}\Psi(x, t)\|_{L^2}$), and also for the case when $g(t, \vec{x}, \Psi(\vec{x}, t))\Psi(\vec{x}, t) = |\Psi(\vec{x}, t)|^\alpha \Psi(\vec{x}, t)$ (for certain α,[1] [25, proposition 7.3.4]):

$$\sup_{t \in [0,\infty]} \|(\vec{x} + it\nabla)\Psi(\vec{x}, t)\|_{L^2} \leq C \tag{7.2.8}$$

We now make a heuristic argument suggesting that this estimate implies the conditions of Proposition 7.2.3. Suppose we have, for some large time, a gaussian at a position $\vec{a}x_0$ (far from the origin, say L units) where $|\vec{a}x_0 - \vec{b}k_0 t| \geq L/2$ (where $L/2$ is chosen simply for concreteness). Then supposing $|\vec{b}k_0|_2 \gg 0$:

$$(\vec{x} + it\nabla)\phi_{(\vec{a},\vec{b})}(\vec{x}) \approx (\vec{a}x_0 - it\vec{b}k_0)\phi_{(\vec{a},\vec{b})}(\vec{x})$$

---

[1] In particular $\alpha \geq (2 - N + \sqrt{N^2 + 12N + 4})/(2N)$.

But then:

$$\left\| (\vec{x} + it\nabla)\phi_{(\vec{a},\vec{b})}(\vec{x}) \right\|_{L^2} \geq (L/2)$$

Therefore, if $\Psi(x, t) = \Psi_{(\vec{a},\vec{b})}(t)\phi_{(\vec{a},\vec{b})}(\vec{x}) + \text{rest}$, then either $\left|\Psi_{(\vec{a},\vec{b})}\right| \leq 2C/L$ or else $\Psi(x, t)$ will violate (7.2.8).

7.3   The Algorithm

We now describe how to construct the approximate solution, $\Phi(\vec{x}, t)$. First, we assume that the various parameters we have described satisfy the constraints given in Sect. 7.4.

The precise mathematical definition of $\Phi(\vec{x}, t)$ is as follows:

$$\Phi(x, nT_{st} + t') = \mathfrak{U}_b(t')\mathcal{P}_{\text{NECC} \cap \text{BB}}\Phi(\vec{x}, nT_{st}) \tag{7.3.1a}$$

$$\Phi(x, (n+1)T_{st}) = \mathfrak{U}_b(T_{st})\mathcal{P}_{\text{NECC} \cap \text{BB}}\Phi(\vec{x}, nT_{st}) \tag{7.3.1b}$$

$$\Phi(x, 0) = \mathcal{P}_{\text{NECC} \cap \text{BB}}\Psi(\vec{x}, 0) \tag{7.3.1c}$$

Here, $0 < t' \leq T_{st}$ and $n \in \mathbb{N}$. Note that $\Phi(\vec{x}, t)$ is not continuous in t at $t = nT_{st}$, due to the filtering.

The critically important part of the algorithm is satisfying the constraints we have described. This ensures that the framelets which we delete from the solution are, in fact, outgoing framelets.

7.4   Choices of the Parameters

There are a number of constraitns on the parameters which need to be satisfied in order for the algorithm to work. One constraint demands that outside the interior box, waves must move freely. That is:

$$L_F \leq L_{in} \tag{7.4.1a}$$

$$\mathfrak{W}^+(\vec{b}, \delta_F, t) \leq \mathfrak{W}^s(\vec{b}, \epsilon, t) \tag{7.4.1b}$$

These are needed by Theorem 7.5.4.

Theorem 7.5.6 imposes a number of conditions on the parameters, nearly all of which are there in order to make sure certain sets of framelets stay inside the box for time T_{st}.

$$\forall \vec{b} \in \text{HF}(k_{\text{inf}}), \ |\vec{b}k_0|_\infty \geq \mathfrak{w}_v^s(\vec{b}, \epsilon) \tag{7.4.1c}$$

$$w \geq 3 \sup_{\vec{b} \in \text{LF}(k_{\text{sup}})} \mathfrak{w}_i^s(\vec{b}, \epsilon) \tag{7.4.1d}$$

$$T_{\text{st}} \leq \frac{w}{3(k_{\text{sup}} + \mathfrak{w}_v^s(\vec{b}, \epsilon))} \tag{7.4.1e}$$

$$L_{\text{in}} \leq L_{\text{inf}} \tag{7.4.1f}$$

$$T_{\text{st}} \leq \inf_{|\vec{b}k_0|_2 \leq k_{\text{sup,NL}}} \frac{L_{\text{in}} + w/2 - L_{\text{NL}}}{k_{\text{sup,NL}} + \mathfrak{w}_v^s(\vec{b}, \epsilon)} \tag{7.4.1g}$$

$$L_{\text{NL}} \leq L_{\text{in}} \tag{7.4.1h}$$

$$\sup_{|\vec{b}k_0|_2 \leq k_{\text{sup,NL}}} \mathfrak{w}_i^s(\vec{b}, \epsilon) \leq w/2 \tag{7.4.1i}$$

$$L_{\text{in}} + w/3 \leq L_{\text{inf}} \tag{7.4.1j}$$

In addition to this list of constraints, one must determine how the parameters k_{sup}, k_{inf}, etc, depend on ϵ, which is dependent on the particular form of the nonlinearity. These can not be determined at this level of generality. For the specific case of linear time-independent potentials, all these relations (except for those involving k_{inf}) are computed in Sect. 8.1, and are listed in (8.1.2).

7.5 Accuracy Estimates

We make some demands on the parameters (L_{in}, T_{st}, etc), which are summarized in Sect. 7.4.

We will first compute the error between $N T_{\text{st}}$ and $(N + 1)T_{\text{st}}$.

Definition 7.5.1 We define the auxiliary functions:

$$\widehat{\mathcal{E}}(t) = \mathcal{A}_F^{-1} \left(\sum_{(\vec{a}, \vec{b}) \in \text{BAD}^C \cap \text{NECC} \cap \text{BB}} \left| \mathfrak{E}_{(\vec{a}, \vec{b})}^s(t) \right|^2 \right)^{1/2} \tag{7.5.1a}$$

$$\widehat{\mathcal{R}}(t) = \mathcal{A}_F^{-1} \left(\sum_{(\vec{a}, \vec{b}) \in \text{NECC}^C \cap \text{BB}} \left| \mathfrak{E}_{(\vec{a}, \vec{b})}^s(t) \right|^2 \right)^{1/2} \tag{7.5.1b}$$

$$\widehat{\mathcal{Q}}(t) = t\mathcal{A}_F^{-1} \left(\sum_{(\vec{a},\vec{b}) \in \mathrm{NL}} \left| \mathfrak{E}_{(\vec{a},\vec{b})}^s(t) \right|^2 \right)^{1/2} \tag{7.5.1c}$$

We now state a simple upper bound on $\widehat{\mathcal{E}}(t)$ and $\widehat{\mathcal{R}}(t)$. In practice, it should not be used. $\widehat{\mathcal{E}}(t)$ and $\widehat{\mathcal{R}}(t)$ are finite sums of known quantities, and thus they should be computed precisely. But it is convenient to demonstrate the order of magnitude of $\widehat{\mathcal{E}}(t)$ and $\widehat{\mathcal{R}}(t)$.

Proposition 7.5.2 *The following inequalities hold for $0 \le t \le T_{\mathrm{st}}$.*

$$\sup_{t \in [0, T_{\mathrm{st}}]} \widehat{\mathcal{E}}(t)$$

$$\le \mathcal{A}_F^{-1} (2L_{\mathrm{WFT}}/x_0)^{N/2} (2k_{\mathrm{sup}}/k_0)^{N/2} \sup_{(\vec{a},\vec{b}) \in \mathrm{BAD}^C \cap \mathrm{NECC} \cap \mathrm{BB}} \left| \mathfrak{E}_{(\vec{a},\vec{b})}^s(t) \right|$$

$$\le \epsilon \mathcal{A}_F^{-1} (2L_{\mathrm{WFT}}/x_0)^{N/2} (2k_{\mathrm{sup}}/k_0)^{N/2} \tag{7.5.2}$$

$$\sup_{t \in [0, T_{\mathfrak{M}}]} \widehat{\mathcal{R}}(t)$$

$$\le \mathcal{A}_F^{-1} (2L_{\mathrm{WFT}}/x_0)^{N/2} (2k_{\mathrm{sup}}/k_0)^{N/2} \sup_{(\vec{a},\vec{b}) \in \mathrm{NECC}^C \cap \mathrm{BB}} \left| \mathfrak{E}_{(\vec{a},\vec{b})}^s(t) \right|$$

$$\le \epsilon \mathcal{A}_F^{-1} (2L_{\mathrm{WFT}}/x_0)^{N/2} (2k_{\mathrm{sup}}/k_0)^{N/2} \tag{7.5.3}$$

$$\sup_{t \in [0, T_{\mathfrak{M}}]} \widehat{\mathcal{Q}}(t)$$

$$\le \mathcal{A}_F^{-1} (2L_{\mathrm{NL}}/x_0)^{N/2} (2k_{\mathrm{sup,NL}}/k_0)^{N/2} \sup_{(\vec{a},\vec{b}) \in \mathrm{NECC}^C \cap \mathrm{BB}} \left| \mathfrak{E}_{(\vec{a},\vec{b})}^s(t) \right|$$

$$\le \epsilon \mathcal{A}_F^{-1} (2L_{\mathrm{NL}}/x_0)^{N/2} (2k_{\mathrm{sup,NL}}/k_0)^{N/2} \tag{7.5.4}$$

Proof A simple calculation. Just count the number of elements in the sums. Then observe that for $t \in [0, T_{\mathrm{st}}]$, $\sup_{(\vec{a},\vec{b}) \in \mathrm{BAD}^C \cap \mathrm{NECC} \cap \mathrm{BB}} \left| \mathfrak{E}_{(\vec{a},\vec{b})}^s(t) \right| \le \epsilon$ by the definition of $\mathrm{BAD} = \mathrm{BAD}(\epsilon, s, T_{\mathrm{st}})$ (and similarly for the other equation).

The bound for $\widehat{\mathcal{Q}}(t)$ is proven by similarly, except by counting the number of elements in L_{NL} (a region in phase space of width L_{NL} in position and $k_{\mathrm{sup,NL}}$ in momentum, see Assumption 4). □

Remark 7.5.3 In practice, Proposition 7.5.2 should not be used. Rather, one can calculate $\widehat{\mathcal{E}}(t)$, $\widehat{\mathcal{R}}(t)$ and $\widehat{\mathcal{Q}}(t)$ precisely. This should be done in practice to choose the parameters. However, we provide these crude upper bounds in order to demonstrate the validity of the method, and to provide rough guidelines as to the choices of the parameters.

7.5.1 Local (1 Step) Error

We first compute the error we make over short time intervals (time $[0, T_{st}]$). We will later string together a number of these short time errors, and calculate the global in time error.

Suppose we are given an initial condition $f(x)$, and an initial error $e(x)$ (the error accumulated from previous timesteps).

We want to compute a bound on:

$$\sup_{t \in [0, T_{st}]} \| \mathfrak{U}(t) f(x) - \mathfrak{U}_b(t) \mathcal{P}_{\text{NECC} \cap \text{BB}}[f(x) + e(x)] \|_{H_b^s} \tag{7.5.5}$$

We first add and subtract $\mathfrak{U}(t|f(x)) \mathcal{P}_{\text{NECC}^C} f(x)$ under the norm, and apply the triangle inequality. Thus, we find:

$$\| \mathfrak{U}(t) f(x) - \mathfrak{U}_b(t) \mathcal{P}_{\text{NECC} \cap \text{BB}}[f(x) + e(x)] \|_{H_b^s}$$

$$\leq \| \mathfrak{U}(t|f) f(x) - \mathfrak{U}(t|f) \mathcal{P}_{\text{NECC}} f(x) \|_{H_b^s}$$

$$+ \| \mathfrak{U}(t|f) \mathcal{P}_{\text{NECC}} f(x) - \mathfrak{U}_b(t) \mathcal{P}_{\text{NECC} \cap \text{BB}}[f(x) + e(x)] \|_{H_b^s} \tag{7.5.6}$$

We state our first result.

Theorem 7.5.4 (Outgoing Waves) *Suppose the following constraints are satisfied:*

$$L_F \leq L_{\text{in}} \tag{7.5.7a}$$

$$\mathfrak{W}^+(\vec{b}, \delta_F, t) \leq \mathfrak{W}^s(\vec{b}, \epsilon, t) \tag{7.5.7b}$$

with $\epsilon(\delta_F)$ defined as in Assumption 5. Then the following holds:

$$\| \mathfrak{U}(t|f) f(x) - \mathfrak{U}(t|f) \mathcal{P}_{\text{NECC}} f(x) \|_{H_b^s} = \| \mathfrak{U}(t|f(x)) \mathcal{P}_{\text{NECC}^C} f(x) \|_{H_b^s}$$

$$\leq \delta_F \| \mathcal{P}_{\text{NECC}^C} f(x) \|_{H_b^s} + \widehat{\mathcal{R}}(t) \| f(x) \|_{L^2}$$

$$+ \mathfrak{H}_+^s(\tilde{g}(\vec{x})) \mathfrak{H}_+^{-s}(e^{-x^2/\sigma^2}) \times \left[\widehat{\mathcal{E}}(T_{st}) \| f(x) + e(x) \|_{L^2} + \right.$$

$$\left. (\widehat{\mathcal{Q}}(T_{st}) \mathcal{G} + t\delta_{\text{NL}}) \sup_{t' \in [0,t]} \| \mathfrak{U}(t')(f(x) + e(x)) \|_{H^s} + \epsilon \right] + \delta_{\text{sup}}$$

$$\equiv \text{OUT}(t) \tag{7.5.8}$$

This is proved in Sect. 10.1 on p. 128. Applying this result, yields:

$$(7.5.6) \leq \text{OUT}(t)$$

$$+ \| \mathfrak{U}(t|f) \mathcal{P}_{\text{NECC}} f(x) - \mathfrak{U}_b(t) \mathcal{P}_{\text{NECC} \cap \text{BB}}[f(x) + e(x)] \|_{H_b^s} \tag{7.5.9}$$

We now add and subtract $\mathfrak{U}(t|\mathcal{P}_{\text{NECC}}f)\mathcal{P}_{\text{NECC}}f(x)$ inside the norm, to obtain:

$$(7.5.9) \leq \text{OUT}(t)$$
$$+ \|\mathfrak{U}(t|f)\mathcal{P}_{\text{NECC}}f(x) - \mathfrak{U}(t|\mathcal{P}_{\text{NECC}}f)\mathcal{P}_{\text{NECC}}f(x)\|_{H_b^s}$$
$$+ \|\mathfrak{U}(t|\mathcal{P}_{\text{NECC}}f)\mathcal{P}_{\text{NECC}}f(x) - \mathfrak{U}_b(t)\mathcal{P}_{\text{NECC}\cap\text{BB}}[f(x)+e(x)]\|_{H_b^s} \qquad (7.5.10)$$

Observing that $\|\mathfrak{U}(t|f)\mathcal{P}_{\text{NECC}}f(x) - \mathfrak{U}(t|\mathcal{P}_{\text{NECC}}f)\mathcal{P}_{\text{NECC}}f(x)\|_{H_b^s}$ is bounded by $\mathcal{L}_{\text{ext}}(t)\,\|\mathcal{P}_{\text{NECC}}f(x)\|_{H_b^s}$ (by Assumption 5), we find:

$$(7.5.10) \leq \text{OUT}(t) + \mathcal{L}_{\text{ext}}(t)\,\|\mathcal{P}_{\text{NECC}}f(x)\|_{H_b^s}$$
$$+ \|\mathfrak{U}(t|\mathcal{P}_{\text{NECC}}f)\mathcal{P}_{\text{NECC}}f(x) - \mathfrak{U}_b(t)\mathcal{P}_{\text{NECC}\cap\text{BB}}[f(x)+e(x)]\|_{H_b^s} \qquad (7.5.11)$$

We add and subtract $\mathfrak{U}(t)\mathcal{P}_{\text{NECC}\cap\text{BB}}\,f(x)$ next, yielding:

$$(7.5.11) \leq \text{OUT}(t) + \mathcal{L}_{\text{ext}}(t)\,\|\mathcal{P}_{\text{NECC}}f(x)\|_{H_b^s}$$
$$+ \|\mathfrak{U}(t|\mathcal{P}_{\text{NECC}}f)\mathcal{P}_{\text{NECC}}f(x) - \mathfrak{U}(t|\mathcal{P}_{\text{NECC}\cap\text{BB}})\mathcal{P}_{\text{NECC}\cap\text{BB}}\,f(x)\|_{H_b^s}$$
$$+ \|\mathfrak{U}(t|\mathcal{P}_{\text{NECC}\cap\text{BB}}f)\mathcal{P}_{\text{NECC}\cap\text{BB}}\,f(x) - \mathfrak{U}_b(t)\mathcal{P}_{\text{NECC}\cap\text{BB}}[f(x)+e(x)]\|_{H_b^s}$$
$$(7.5.12)$$

We state another result:

Theorem 7.5.5 (Residual Waves) *The residual waves satisfy the following estimate:*

$$\|\mathfrak{U}(t|\mathcal{P}_{\text{NECC}}f)\mathcal{P}_{\text{NECC}}f(x) - \mathfrak{U}(t|\mathcal{P}_{\text{NECC}\cap\text{BB}})\mathcal{P}_{\text{NECC}\cap\text{BB}}\,f(x)\|_{H_b^s}$$
$$\leq \mathcal{L}(t)\Bigg(\mathfrak{H}_+^s(\tilde{g}(\vec{x}))\mathfrak{H}_+^{-s}(e^{-x^2/\sigma^2})\Big[\widehat{\mathcal{E}}(T_{\text{st}})\,\|f(x)+e(x)\|_{L^2}$$
$$+ (\widehat{\mathcal{Q}}(T_{\text{st}})\mathcal{G} + t\delta_{\text{NL}})\sup_{t'\in[0,t]}\big\|\mathfrak{U}(t')(f(x)+e(x))\big\|_{H^s} + \epsilon\Big]$$
$$+ \delta_{\text{inf}}\Bigg) \equiv \text{RES}(t) \qquad (7.5.13)$$

This is proved in Sect. 10.2 on p. 129.
   Applying this yields:

$$(7.5.12) \leq \text{OUT}(t) + \mathcal{L}_{\text{ext}}(t)\,\|\mathcal{P}_{\text{NECC}}f(x)\|_{H_b^s}$$
$$+ \text{RES}(t)$$
$$+ \|\mathfrak{U}(t)\mathcal{P}_{\text{NECC}\cap\text{BB}}\,f(x) - \mathfrak{U}_b(t)\mathcal{P}_{\text{NECC}\cap\text{BB}}[f(x)+e(x)]\|_{H_b^s} \qquad (7.5.14)$$

We now add and subtract $\mathfrak{U}(t|\mathcal{P}_{\text{NECC}\cap\text{BB}}(f+e))\mathcal{P}_{\text{NECC}\cap\text{BB}}(f(x)+e(x))$, and bound this by $\mathcal{L}(t)\,\|e(x)\|_{H_b^s}$:

$$(7.5.14) \leq \text{OUT}(t) + \mathcal{L}_{\text{ext}}(t)\,\|\mathcal{P}_{\text{NECC}}f(x)\|_{H_b^s}$$

$$+ \text{RES}(t)$$

$$+ \|e(x)\|_{H_b^s}\,\mathcal{L}(t)$$

$$+ \|\mathfrak{U}(t)\mathcal{P}_{\text{NECC}\cap\text{BB}}[f(x)+e(x)] - \mathfrak{U}_b(t)\mathcal{P}_{\text{NECC}\cap\text{BB}}[f(x)+e(x)]\|_{H_b^s}$$
$$(7.5.15)$$

Finally, we bound the last term as follows.

Theorem 7.5.6 (Lingering Waves) *Let the nonlinearity satisfy Assumption 3. Let the following conditions on the parameters be satisfied:*

$$\forall \vec{b} \in \text{HF}(k_{\text{inf}}), \ |\vec{b}k_0|_\infty \geq \mathfrak{w}_v^s(\vec{b},\epsilon) \tag{7.5.16a}$$

$$w \geq 3 \sup_{\vec{b}\in\text{LF}(k_{\text{sup}})} \mathfrak{w}_i^s(\vec{b},\epsilon) \tag{7.5.16b}$$

$$T_{\text{st}} \leq \frac{w}{3(k_{\text{sup}} + \mathfrak{w}_v^s(\vec{b},\epsilon))} \tag{7.5.16c}$$

$$L_{\text{in}} \leq L_{\text{inf}} \tag{7.5.16d}$$

$$T_{\text{st}} \leq \inf_{|\vec{b}k_0|_2 \leq k_{\text{sup,NL}}} \frac{L_{\text{in}} + w/2 - L_{\text{NL}}}{k_{\text{sup,NL}} + \mathfrak{w}_v^s(\vec{b},\epsilon)} \tag{7.5.16e}$$

$$L_{\text{NL}} \leq L_{\text{in}} \tag{7.5.16f}$$

$$\sup_{|\vec{b}k_0|_2 \leq k_{\text{sup,NL}}} \mathfrak{w}_i^s(\vec{b},\epsilon) \leq w/2 \tag{7.5.16g}$$

$$L_{\text{in}} + w/3 \leq L_{\text{inf}} \tag{7.5.16h}$$

Let $\Psi(x, t = 0) = \mathcal{P}_{\text{NECC}\cap\text{BB}}\Psi_0(x)$. Then the following estimate holds:

$$\|(\mathfrak{U}(t) - \mathfrak{U}_b(t))\mathcal{P}_{\text{NECC}\cap\text{BB}}\Psi_0(x)\|_{H_b^s}$$

$$\leq (E(t) + Q(t)) + \mathcal{G}e^{\mathcal{G}t} \star (E(t) + Q(t)) \tag{7.5.17a}$$

$$\|(\mathfrak{U}(t) - \mathfrak{U}_b(t))\mathcal{P}_{\text{NECC}\cap\text{BB}}\Psi_0(x)\|_{H_b^s}$$

$$\leq (E(t) + Q_b(t)) + \mathcal{G}e^{\mathcal{G}t} \star (E(t) + Q_b(t)) \tag{7.5.17b}$$

The free error and interaction error are given by:

$$E(t) \leq \widehat{\mathcal{E}}(t) \, \|\Psi\|_{L^2} + 2\delta_{\inf} \tag{7.5.18a}$$

$$Q(t) \leq (\widehat{\mathcal{Q}}(t)\mathcal{G} + t\delta_{\mathrm{NL}}) \, \|\Psi\|_{H^s} \tag{7.5.18b}$$

A similar estimate holds for $Q_b(t)$ but with $\Psi(\vec{x}, t)$ replaced by $\Psi_b(\vec{x}, t)$. The functions $\widehat{\mathcal{E}}(t)$ and $\widehat{\mathcal{Q}}(t)$ are defined in 7.5.1 on p. 80.

This result is proved in Sect. 9. Applying this result shows that:

$$(7.5.15) \leq \mathrm{OUT}(t) + \mathcal{L}_{\mathrm{ext}}(t) \, \|\mathcal{P}_{\mathrm{NECC}} f(x)\|_{H^s_b} + \mathrm{RES}(t)$$

$$+ \, \|e(x)\|_{H^s_b} \mathcal{L}(t)$$

$$+ \, (1 + \mathcal{G}e^{\mathcal{G}t}\star)(\widehat{\mathcal{E}}(t) \, \|\Psi\|_{L^2} + 2\delta_{\inf} + (\widehat{\mathcal{Q}}(t)\mathcal{G} + t\delta_{\mathrm{NL}}) \, \|\Psi\|_{H^s}) \tag{7.5.19}$$

Remark 7.5.7 This analysis can be extended to encompass discretization errors. Assuming one has control of discretization errors on the box, one can simply include these errors in $e(x)$. We do not do this here, since it is well beyond the scope of this paper.

7.5.2 Global Error Estimates

Given the above result on the one timestep error, we now compute the global-in-time error.

At time $t = 0$, we let $f(x) = \Psi_0(x)$ and $e(x) = 0$. At time nT_{st} $(n = 1, \dots, N)$, we let $f(x) = \Phi(x, nT_{\mathrm{st}})$ and

$$e(x) = \mathfrak{U}(T_{\mathrm{st}})\mathcal{P}_{\mathrm{NECC}\cap\mathrm{BB}}\Phi(x, (n-1)T_{\mathrm{st}})$$

$$- \mathfrak{U}_b(T_{\mathrm{st}})\mathcal{P}_{\mathrm{NECC}\cap\mathrm{BB}}\Phi(x, (n-1)T_{\mathrm{st}})$$

Putting this all together, for $n = 0 \dots M$, with $M = T_{\mathfrak{M}}/T_{\mathrm{st}}$, we find:

$$\|\mathfrak{U}(MT_{\mathrm{st}})\Psi_0(x) - \Phi(x, MT_{\mathrm{st}})\|_{H^s_b}$$

$$\leq \sum_{n=0}^{M} \Bigg(\mathrm{OUT}((M-n)T_{\mathrm{st}}) + \mathcal{L}_{\mathrm{ext}}((M-n)T_{\mathrm{st}}) \, \|\mathcal{P}_{\mathrm{NECC}} f(x)\|_{H^s_b}$$

$$+ \mathcal{L}((M-n)T_{\mathrm{st}}) \, \mathrm{RES}(nT_{\mathrm{st}}) + \mathcal{L}((M-n)T_{\mathrm{st}}) \, \mathrm{BoxError}(nT_{\mathrm{st}}) \Bigg) \tag{7.5.20}$$

The term $\mathrm{BoxError}(nT_{\mathrm{st}})$ is bounded by (7.5.17) with $\Psi_0(x) = \Phi(x, nT_{\mathrm{st}})$.

Remark 7.5.8 This result is essentially what one would expect. The term RES(nT_{st}) represents the main source of error. This is the error caused by waves we cannot filter with our algorithm. The error bound says that at tine nT_{st}, we make an error of size RES(nT_{st}). After that, the error grows at a rate $\mathcal{L}(t - nT_{st})$.

BoxError(nT_{st}) represents the error due to filtering at time nT_{st}, and this also grows at the rate $\mathcal{L}(t - nT_{st})$ after that.

We now wish to make sense of (7.5.20). We first substitute everything in order to get a complete picture. We will then rearrange and simplify significantly.

$$\|\mathfrak{U}(NT_{st})\Psi_0(x) - \Phi(x, NT_{st})\|_{H_b^s}$$

$$\leq \sum_{n=0}^{M} \delta_F \left\|\mathcal{P}_{\text{NECC}^c}\Phi(x, nT_{st})\right\|_{H_b^s} + \widehat{\mathcal{R}}((M-n)T_{st}) \left\|\Phi(x, nT_{st})\right\|_{L^2}$$

$$+ \mathfrak{H}_+^s(\tilde{g}(\vec{x}))\mathfrak{H}_+^{-s}(e^{-x^2/\sigma^2})\left[\widehat{\mathcal{E}}(T_{st}) \left\|\Phi(x, (n-1)T_{st})\right\|_{L^2}\right.$$

$$+ (\widehat{\mathcal{Q}}(T_{st}) + t\delta_{NL}) \sup_{t'\in[0,t]} \left\|\mathfrak{U}(t')\Phi(x, (n-1)T_{st})\right\|_{H^s} + \epsilon\left] + \delta_{\sup}\right.$$

$$+ \mathcal{L}_{\text{ext}}((M-n)T_{st}) \|\mathcal{P}_{\text{NECC}}\Phi(x, nT_{st})\|_{H^s}$$

$$\mathcal{L}((M-n)T_{st})\left(\mathfrak{H}_+^s(\tilde{g}(\vec{x}))\mathfrak{H}_+^{-s}(e^{-x^2/\sigma^2})\right.$$

$$\left[\widehat{\mathcal{E}}(T_{st}) \|\Phi(x, (n-1)T_{st})\|_{L^2}\right.$$

$$+ (\widehat{\mathcal{Q}}(T_{st}) + t\delta_{NL}) \sup_{t'\in[0,t]} \left\|\mathfrak{U}(t')\Phi(x, (n-1)T_{st})\right\|_{H^s} + \epsilon\left] + \delta_{\inf}\right)$$

$$\mathcal{L}((M-n)T_{st})(1 + T_{st}\mathcal{G}e^{\mathcal{G}T_{st}})\left[\widehat{\mathcal{E}}(T_{st}) \|\Phi(x, nT_{st})\|_{L^2} + \delta_{\inf}\right.$$

$$\left.+ (\widehat{\mathcal{Q}}(T_{st}) + T_{st}\delta_{NL}) \sup_{t\in[0,T_{st}]} \|\Phi(x, nT_{st}+t)\|_{L^2}\right] \qquad (7.5.21)$$

We now take this and collect all the terms containing

$$\left[\widehat{\mathcal{E}}(T_{st}) \|\Phi(x, nT_{st})\|_{L^2} + \widehat{\mathcal{Q}}(T_{st}) \sup_{t\in[0,T_{st}]} \|\Phi(x, nT_{st}+t)\|_{L^2}\right]$$

as well as $\delta_{\inf}$, δ_F, $\mathcal{L}_{\text{ext}}(t)$, etc. We also replace the terms $\|\mathcal{P}_F\Phi(\vec{x}, t)\|_{L^2}$ by $\sqrt{\mathcal{B}_F/\mathcal{A}_F} \|\Phi(\vec{x}, t)\|_{L^2}$ and $\|\mathcal{P}_F\Phi(\vec{x}, t)\|_{H^s}$ by $\mathfrak{H}_+^s(\tilde{g}(\vec{x}))\mathfrak{H}_+^{-s}(e^{-x^2/\sigma^2})\|\Phi(\vec{x}, t)\|_{H^s}$. We thus arrive at our main theorem.

Theorem 7.5.9 (Global Error Bound) *We have the following bound on the error:*

$$
\sup_{t\in[0,T_{\mathfrak{M}}]} \|\mathfrak{U}(t)\Psi_0(x) - \Phi(x,t)\|_{H_b^s} \leq (7.5.21) \leq \sup_{t\in[0,T_{\mathfrak{M}}]} \|\Phi(\vec{x},t)\|_{H^s} \Bigg[
$$

$$
\widehat{\mathcal{E}}(T_{\text{st}}) \left(\left[1 + T_{\text{st}}\mathcal{G}e^{\mathcal{G}T_{\text{st}}} + 2\mathfrak{H}_+^s(\tilde{g}(\vec{x}))\mathfrak{H}_+^{-s}(e^{-\frac{x^2}{\sigma^2}})\right] \sum_{n=0}^{M} \mathcal{L}((M-n)T_{\text{st}}) \right)
$$

$$
+ \left(\sum_{n=0}^{M} \widehat{\mathcal{R}}((M-n)T_{\text{st}}) \right)
$$

$$
+ \widehat{\mathcal{Q}}(T_{\text{st}}) \left[(2 + T_{\text{st}}\mathcal{G}e^{\mathcal{G}T_{\text{st}}}) \left(\sum_{n=0}^{M} \mathcal{L}((M-n)T_{\text{st}}) \right) \right.
$$

$$
\left. + (T_{\mathfrak{M}}/T_{\text{st}})\mathfrak{H}_+^s(\tilde{g}(\vec{x}))\mathfrak{H}_+^{-s}(e^{-x^2/\sigma^2}) \right]
$$

$$
+ \delta_{\text{NL}}T_{\text{st}} \left(\sum_{n=0}^{M} \widehat{\mathcal{R}}((M-n)T_{\text{st}}) \right) (2 + T_{\text{st}}\mathcal{G}e^{\mathcal{G}T_{\text{st}}})
$$

$$
+ \delta_F(T_{\mathfrak{M}}/T_{\text{st}}) + \left(\sum_{n=0}^{M} \mathcal{L}_{\text{ext}}((M-n)T_{\text{st}}) \right)
$$

$$
\Bigg]
$$

$$
+ \delta_{\text{inf}}(2 + T_{\text{st}}\mathcal{G}e^{\mathcal{G}T_{\text{st}}}) \left(\sum_{n=0}^{M} \mathcal{L}((M-n)T_{\text{st}}) \right) + \delta_{\text{sup}}(T_{\mathfrak{M}}/T_{\text{st}}) \qquad (7.5.22)
$$

Although the error bound looks complicated, each term has a simple meaning.

The term $\delta_{\text{NL}}(\dots)$ is similar. This term measures how much of the nonlinearity actually outside the computational domain. In order to accurately compute the effect of the nonlinearity, it must be contained inside the computational domain. Thus, whatever mass exists outside the computational region (namely $[-L_{\text{NL}}, L_{\text{NL}}]^N \times [-k_{\text{sup,NL}}, k_{\text{sup,NL}}]^N$ in phase space) will also cause an error.

The term $\delta_F(T_{\mathfrak{M}}/T_{\text{st}})$ measures how much of the solution which we thought was outgoing actually wasn't. That is, we examined each gaussian, and determined that under the free flow, that particular gaussian was leaving the computational domain. But although the flow is nearly free on the boundary, it is possible that some small fraction of the waves we believe are outgoing are returning. That is measured by δ_F.

The next piece,

$$\left(\sum_{n=0}^{M} \mathcal{L}_{\text{ext}}((M-n)T_{\text{st}}) \right),$$

is a little bit trickier to describe. This part of the error measures how the nonlinearity changes in response to the small errors made when we filter off the outgoing waves. In the event the "nonlinearity" is linear, this term is identically zero. But in other cases, it may grow rather large with t.

It is best illustrated by an example. Consider the NLS:

$$i\partial_t \Psi(x,t) = (-(1/2)\Delta_b + V(x))\Psi(x,t) + f(|\Psi(x,t)|^2)\Psi(x,t)$$

with $V(x)$ an even, real valued potential having two (nonlinear) bound states, and $(|\Psi(x,t)|)$ a monotone real-positive function satisfying certain other constraints (see [64]). It is known that this system exhibits ground state selection [64]. That is, if $\Psi(x,0)$ is an odd function, then $\Psi(x,t)$ remains situated on the odd (excited) bound state for all time. If, however, we replace $\Psi(x,0) = \text{odd}(x) + \epsilon\,\text{even}(x)$, then half the mass of $\Psi(x,t)$ will radiate off to infinity, while the other half will be trapped in the ground state.

The function $\mathcal{L}_{\text{ext}}(t)$ measures the capacity of the system to behave nonlinearly in response to perturbations, in a manner like that which we just described.

The last term, $\delta_{\text{sup}}(T_{\mathfrak{M}}/T_{\text{st}})$, is essentially the amount of mass at frequencies higher than k_{sup}. Although k_{sup} (as used in Assumption 2) is slightly different from the usual definition of k_{sup} (namely $k_{\text{sup}} = \pi/\Delta x$, with Δx the lattice spacing in position on the grid), it is a very similar object, namely the largest frequency we can resolve.

Finally, we come to the term containing $k_{\text{inf}}(\ldots)$. This term contains waves with frequency sufficiently low so that it is very difficult to tell if they are entering the box or leaving. This is basically due to the fact that for functions localized in the filter region, the Heisenberg uncertainty principle says that we cannot determine whether low frequency waves are incoming or outgoing. In most of our experiments, this was the dominant term in the error.

We now prove a corollary to Theorem 7.5.9, which states that under the assumptions given in Sect. 7.1, we can make the error due to boundary reflections vanish by making certain explicit choices of the parameters.

Corollary 7.5.10 (Convergence to Zero) *We can choose the parameters T_{st}, L_{in} and w in such a way that for any $\tau > 0$ and $T_{\mathfrak{M}} < \infty$,*

$$\sup_{t \in [0, T_{\mathfrak{M}}]} \|\mathfrak{U}(t)\Psi_0(x) - \Phi(x,t)\|_{H_b^s} \leq \tau \tag{7.5.23}$$

The proof is deferred to the end of this section, after we have discussed the sources of the error.

Proof of Corollary 7.5.10 We show here how we can make the error bound (7.5.22) arbitrarily small.

We begin by considering the terms δ_{inf}, δ_{sup} δ_{NL}, δ_F and $\mathcal{L}_{ext}(t)$ found in the last four lines of (7.5.22). According to Assumptions 6, 2, 4 and 5 (respectively), we can choose the parameters k_{inf}, k_{sup}, L_{NL}, $k_{sup,NL}$ and L_F in such a way that δ_{inf}, δ_{sup} δ_{NL}, δ_F and $\mathcal{L}_{ext}(t)^2$ are all arbitrarily small. Therefore, it is possible to choose these parameters in such a way that:

$$
\delta_{NL} \sup_{t \in [0,T_{\mathfrak{M}}]} \|\Phi(\vec{x},t)\|_{H^s} T_{st} \left(\sum_{n=0}^{M} \widehat{\mathcal{R}}((M-n)T_{st}) \right) (2 + T_{st} \mathcal{G} e^{\mathcal{G}T_{st}})
$$

$$
+ \delta_F \sup_{t \in [0,T_{\mathfrak{M}}]} \|\Phi(\vec{x},t)\|_{H^s} (T_{\mathfrak{M}}/T_{st})
$$

$$
+ \left(\sum_{n=0}^{M} \mathcal{L}_{ext}((M-n)T_{st}) \right) \sup_{t \in [0,T_{\mathfrak{M}}]} \|\Phi(\vec{x},t)\|_{H^s}
$$

$$
+ \delta_{inf}(2 + T_{st} \mathcal{G} e^{\mathcal{G}T_{st}}) \left(\sum_{n=0}^{M} \mathcal{L}((M-n)T_{st}) \right) + \delta_{sup}(T_{\mathfrak{M}}/T_{st})
$$

$$
\leq \tau/2 \qquad (7.5.24)
$$

The exact manner in which this will be done is highly model dependent. Later on (in Remark 7.5.11) we will discuss briefly the obvious way to make this small, and why this may not be the best way to satisfy (7.5.24).

We now take k_{inf}, k_{sup}, L_{NL}, $k_{sup,NL}$ and L_F to be fixed quantities.

Once these terms are chosen, we must choose $L_{in,w}$ satisfying the various constraints. After this is done, Proposition 7.5.2 provides a bound on $\widehat{\mathcal{E}}(T_{st})$, $\widehat{\mathcal{R}}(T_{st})$ and $\widehat{\mathcal{Q}}(T_{st})$—in particular each is bounded by *const* $\times \epsilon$ (with *const* a function of the various parameters).

More precisely, we do the following. We now need to obtain the following bound:

$$
\sup_{t \in [0,T_{\mathfrak{M}}]} \|\Phi(\vec{x},t)\|_{H^s} \left[\widehat{\mathcal{E}}(T_{st}) \right.
$$

$$
\times \left(\left[(1 + T_{st} \mathcal{G} e^{\mathcal{G}T_{st}}) + 2\mathfrak{H}^s_+(\tilde{g}(\vec{x}))\mathfrak{H}^{-s}_+(e^{-x^2/\sigma^2}) \right] \sum_{n=0}^{M} \mathcal{L}((M-n)T_{st}) \right)
$$

$$
\left. + \left(\sum_{n=0}^{M} \widehat{\mathcal{R}}((M-n)T_{st}) \right) \right.
$$

---

[2] Recall that $\sup_{t \in [0,T_{\mathfrak{M}}]} \mathcal{L}_{ext}(t) \leq \delta_F$. In principle, one could simply use this bound. However, in practice, we expect that $\sum_{n=0}^{M} \mathcal{L}_{ext}((M-n)T_{st}) \ll M\delta_F$, so this would be an inefficient choice.

$$+ \widehat{\mathcal{Q}}(T_{\text{st}}) \left[(2 + T_{\text{st}} \mathcal{G} e^{\mathcal{G} T_{\text{st}}}) \left(\sum_{n=0}^{M} \mathcal{L}((M - n) T_{\text{st}}) \right) \right.$$

$$\left. + (T_{\mathfrak{M}} / T_{\text{st}}) \mathfrak{H}_+^s (\tilde{g}(\vec{x})) \mathfrak{H}_+^{-s} (e^{-x^2/\sigma^2}) \right] \right] \leq \tau/2 \qquad (7.5.25)$$

We recall the bounds computed in Proposition 7.5.2, and substitute them in to obtain:

$$(7.5.25) \leq \sup_{t \in [0, T_{\mathfrak{M}}]} \| \Phi(\vec{x}, t) \|_{H^s} \left[\vphantom{\sum} \right.$$

$$\epsilon \mathcal{A}_F^{-1} (2 L_{\text{WFT}} / x_0)^{N/2} (2 k_{\text{sup}} / k_0)^{N/2}$$

$$\times \left(\left[(1 + T_{\text{st}} \mathcal{G} e^{\mathcal{G} T_{\text{st}}}) + 2 \mathfrak{H}_+^s (\tilde{g}(\vec{x})) \mathfrak{H}_+^{-s} (e^{-x^2/\sigma^2}) \right] \sum_{n=0}^{M} \mathcal{L}((M - n) T_{\text{st}}) \right)$$

$$+ \epsilon \mathcal{A}_F^{-1} (2 L_{\text{WFT}} / x_0)^{N/2} (2 k_{\text{sup}} / k_0)^{N/2} (T_{\mathfrak{M}} / T_{\text{st}})$$

$$+ \epsilon \mathcal{A}_F^{-1} (2 L_{\text{NL}} / x_0)^{N/2} (2 k_{\text{sup,NL}} / k_0)^{N/2}$$

$$\times \left[(2 + T_{\text{st}} \mathcal{G} e^{\mathcal{G} T_{\text{st}}}) \left(\sum_{n=0}^{M} \mathcal{L}((M - n) T_{\text{st}}) \right) \right.$$

$$\left. \left. + (T_{\mathfrak{M}} / T_{\text{st}}) \mathfrak{H}_+^s (\tilde{g}(\vec{x})) \mathfrak{H}_+^{-s} (e^{-x^2/\sigma^2}) \right] \right] \qquad (7.5.26)$$

We observe that this is linear in ϵ. Thus, by making the choice:

$$\epsilon^{-1} = 2\tau^{-1} \sup_{t \in [0, T_{\mathfrak{M}}]} \| \Phi(\vec{x}, t) \|_{H^s} \left[\vphantom{\sum} \right.$$

$$\mathcal{A}_F^{-1} (2 L_{\text{WFT}} / x_0)^{N/2} (2 k_{\text{sup}} / k_0)^{N/2}$$

$$\times \left(\left[(1 + T_{\text{st}} \mathcal{G} e^{\mathcal{G} T_{\text{st}}}) + 2 \mathfrak{H}_+^s (\tilde{g}(\vec{x})) \mathfrak{H}_+^{-s} (e^{-x^2/\sigma^2}) \right] \sum_{n=0}^{M} \mathcal{L}((M - n) T_{\text{st}}) \right)$$

$$+ \mathcal{A}_F^{-1} (2 L_{\text{WFT}} / x_0)^{N/2} (2 k_{\text{sup}} / k_0)^{N/2} (T_{\mathfrak{M}} / T_{\text{st}})$$

$$+ \mathcal{A}_F^{-1} (2 L_{\text{NL}} / x_0)^{N/2} (2 k_{\text{sup,NL}} / k_0)^{N/2}$$

$$\times \left[(2 + T_{\text{st}} \mathcal{G} e^{\mathcal{G} T_{\text{st}}}) \left(\sum_{n=0}^{M} \mathcal{L}((M - n) T_{\text{st}}) \right) \right.$$

$$\left. \left. + (T_{\mathfrak{M}} / T_{\text{st}}) \mathfrak{H}_+^s (\tilde{g}(\vec{x})) \mathfrak{H}_+^{-s} (e^{-x^2/\sigma^2}) \right] \right]$$

we find that (7.5.25) $\leq$ (7.5.26) $\leq \tau/2$. This holds only of $\epsilon \leq \epsilon(\delta_F)$, with $\epsilon(\delta_F)$ defined in Assumption 5.

Thus, by this choice of parameters, we have made the error smaller than τ. $\qquad \square$

Remark 7.5.11 The obvious way to make (7.5.24) small is to make the following choices for $k_{\inf}$, $k_{\sup}$, L_{NL}, $k_{\sup,NL}$ and L_F:

$$L_{NL} = L_{NL} \left(\frac{\tau}{10 \sup_{t \in [0, T_{\mathfrak{M}}]} \| \Phi(\vec{x}, t) \|_{H^s}} \right.$$

$$\left. \times \frac{1}{T_{st} \left(\sum_{n=0}^{M} \widehat{\mathcal{R}}((M-n)T_{st}) \right) (2 + T_{st}\mathcal{G}e^{\mathcal{G}T_{st}})} \right) \qquad (7.5.27a)$$

$$k_{\sup,NL} = k_{\sup,NL} \left(\frac{\tau}{10 \sup_{t \in [0, T_{\mathfrak{M}}]} \| \Phi(\vec{x}, t) \|_{H^s}} \right.$$

$$\left. \times \frac{1}{T_{st} \left(\sum_{n=0}^{M} \widehat{\mathcal{R}}((M-n)T_{st}) \right) (2 + T_{st}\mathcal{G}e^{\mathcal{G}T_{st}})} \right) \qquad (7.5.27b)$$

$$L_F = L_F \left(\frac{\tau T_{st}}{20 T_{\mathfrak{M}} \sup_{t \in [0, T_{\mathfrak{M}}]} \| \Phi(\vec{x}, t) \|_{H^s}} \right) \qquad (7.5.27c)$$

$$k_{\inf} = k_{\inf} \left(\frac{\tau}{10(2 + T_{st}\mathcal{G}e^{\mathcal{G}T_{st}}) \left(\sum_{n=0}^{M} \mathcal{L}((M-n)T_{st}) \right)} \right) \qquad (7.5.27d)$$

$$k_{\sup} = k_{\sup} \left(\frac{\tau T_{st}}{10 T_{\mathfrak{M}}} \right) \qquad (7.5.27e)$$

This particular choice ensures that each term on the right hand side of (7.5.24) is smaller than $\tau/10$. Since there are 5 terms on the right, the whole thing is less than $\tau/2$.

Although obvious and clearly effective, this choice is likely to be inefficient. Supposing one term to be significantly more expensive than the others (e.g. one term being polynomial in τ^{-1}, the rest being logarithmic), it makes more sense to make the expensive term only smaller than, e.g. $(1 - \delta)\tau/2$, and make each of the others smaller than $\delta\tau/2$ (with $\delta \ll 1/2$).

Thus, although we illustrate that this can be done with (7.5.27), we emphasize that the exact method of satisfying (7.5.24) is strongly dependent on the particular model chosen.

Remark 7.5.12 To get from (7.5.25) to (7.5.26), we made use of the the weak form of Proposition 7.5.2. That is to say, in the bounds on $\widehat{\mathcal{E}}(t)$, $\widehat{\mathcal{R}}(t)$ and $\widehat{\mathcal{Q}}(t)$, we had an intermediate estimate which appeared unwieldy. Nevertheless, the intermediate estimate is far sharper, and is the one that should be used in practice. We used the less sharp estimate simply to demonstrate that $\widehat{\mathcal{E}}(t)$, $\widehat{\mathcal{R}}(t)$ and $\widehat{\mathcal{Q}}(t)$ are quantities which we can make arbitrarily small.

7.6 Remarks

7.6.1 Almost Optimality of the Estimates

The estimates we give here are crude at some points, and can probably be improved significantly. However, in principle, we believe that a result of the form (7.5.20) is the best possible result one can hope for with our method, or any other method based on time stepping.

The reason for this is the following. Consider any numerical method based on time integration. Suppose that it makes an error (however small) at times t_0. This error has now been made, and it is highly unlikely that further errors will completely cancel it. Suppose after t_0, we have the ability to propagate further with no error (but we need to take the incorrect result $\Phi(x, t_0)$ as an initial condition). Then $\|\mathfrak{U}(t)\Psi(x, t_0) - \mathfrak{U}(t)\Phi(x, t_0)\|_{H_b^s}$ is only bounded by $\mathcal{L}(t) \|\Psi(x, t_0) - \Phi(x, t_0)\|_{H^s}$. Repeating this argument every time an error is made leads to a bound very similar to ours.

7.6.2 Difference with the Dirichlet-to-Neumann Approach

Unlike the Dirichlet-to-Neumann approach, the TDPSF is not embedded in a hierarchy of increasingly accurate boundary conditions. The reason for this is that we are not attempting to construct the exact solution on the boundary. Rather, we are merely assuming the wave behaves freely and semiclassically on the boundary, and filtering it based on this. Thus, apart from increasing w (and hence making this assumption more accurate), we have little recourse to increase the accuracy of this method. So although our method is highly accurate, we can not increase the accuracy without bound while leaving the size of the box fixed.

7.6.3 The Ping Pong Phenomenon

One potential difficulty in solving time dependent problems is that a problem which is stable on $\mathbb{R}^N$ may exhibit long time instability on a periodic boxes. Given a box $[-L_{\text{trunc}}, L_{\text{trunc}}]^N$ with periodic boundaries, Bourgain (c.f. [21]) has proven the existence of a time dependent potential $V(\vec{x}, t)$ which is smooth and well localized in $\vec{x}$ having the property that $\|\mathfrak{U}_b(t)\Psi_0\|_{H_b^s}$ grows logarithmically in time. This occurs because the time dependent potential essentially plays a quantum mechanical variant of "ping pong".

This suggests that some numerical methods might exhibit this long time instability if one attempt to solve (1.1.1) on $\mathbb{R}^N$ with such a potential. However, our method prevents this from occurring. We do this by periodically removing all framelets which move faster than k_{sup} (since we filter off waves which are outside of NECC ∩ BB, and BB has no framelets with $\vec{b}k_0 \geq k_{sup}$).

7.6.4 Bounds on k_{inf}

Another potential difficulty comes from the fact that in general, one has no bounds on k_{inf}. We describe here a situation with a simple linear (time-dependent) potential for which k_{inf} can be arbitrarily small while leaving the potential bounded and smooth in any reasonable norm.

Consider a nonlinearity of the form $g(t, \vec{x}, \Psi(\vec{x}, t))\Psi(\vec{x}, t) = V(\vec{x}, t)\Psi(\vec{x}, t)$. We suppose that $V(\vec{x}, t)$ takes the form $V_0(x - (e/\omega^2)\cos(\omega t))$ for some smooth, rapidly decaying potential $V_0(x)$.

This system is equivalent, by a unitary gauge transformation, to the time dependent system with Hamiltonian $H(t) = -(1/2)\Delta + V_0(x) + e\cos(\omega t) \cdot x$ (c.f. [26], chapter 7).

Now, suppose further that the reference Hamiltonian $H_0 = -(1/2)\Delta + V_0(x)$ has a single bound state, having energy $-E_0$.

Consider an initial condition initially localized in this bound state.

In this case, Fermi's golden rule suggests that for e small and $\omega > |E_0|$, mass will be ejected from the bound state into the continuum,[3] and will have energy $\omega - E_0$ after ejection. Thus, energy transitions from the bound state into frequencies localized near $\sqrt{\omega - E_0}$. By making ω sufficiently close to E_0, we can make this as small as possible.

Thus, in this scenario, $k_{inf} \ll \sqrt{\omega - E_0}$, i.e. k_{inf} can be made as small as desired.

Examples of this form can be dealt with by a multiscale extension of our method, as described in [63]. The computational complexity in this case grows logarithmically in k_{inf}, or of $k_{inf} = 0$, logarithmically in $T_{\mathfrak{M}}$.

---

[3] This happens only generically. More precisely, it happens if $\langle u_0(x)|e \cdot xu(x, \omega - E_0)\rangle \neq 0$, where $u_0(x)$ is the bound state and $u(x, \omega - E_0)$ is the generalized eigenfunction at energy $\omega - E_0$.

Chapter 8
Discussions on the Assumptions

In this part we verify that the assumptions hold for a common example.

8.1 Stationary Potentials

In this section, we consider the case where $g(t, \vec{x}, \Psi(\vec{x}, t))\Psi(\vec{x}, t)$ is merely a time independent linear potential. That is, $g(t, \vec{x}, \Psi(\vec{x}, t))\Psi(\vec{x}, t) = V(x)\Psi(\vec{x}, t)$ for $V(x)$ an analytic, real valued finite range potential. More precisely, we demand the following:

$$\left| \partial_x^j V(x) \right| \le C_V \langle x \rangle^{-(1+\beta)} \tag{8.1.1a}$$

$$\hat{V}(k) \le C_V' e^{-\alpha|\vec{k}|} \tag{8.1.1b}$$

We make further restrictions; we assume $\Psi(x, 0) \in H^1(\mathbb{R}^N)$ (finite kinetic energy), and we assume $\|\Psi(x, 0)\|_{L^2} = 1$ (normalization). As a convention, let $E_0 = \langle \Psi(x, 0)|H\Psi(x, 0)\rangle$ denote the energy of the system (a conserved quantity).

For this case, we are able to validate Assumptions 1–5. We obtain the following bounds on the constants (which we will prove shortly):

$$\mathcal{L}(t) = 1 \tag{8.1.2a}$$

$$\mathcal{L}_{\text{ext}}(t) = 0 \tag{8.1.2b}$$

© The Author(s), under exclusive license to Springer Nature Singapore Pte Ltd. 2023
A. Soffer et al., *Time Dependent Phase Space Filters*, SpringerBriefs on PDEs and Data Science, https://doi.org/10.1007/978-981-19-6818-1_8

$$k_{\text{sup}} = \mathfrak{K}^s(k_{\text{sup}}) + \delta_{\text{sup}}^{-1}\sqrt{|E_0| + \|\Psi(x,0)\|_{L^2}^2 (\|V(x)\|_{L^\infty} + 1/2)}$$

$$\times 2^{3/2}\mathfrak{H}_+^s(\tilde{g}(\vec{x}))\mathfrak{H}_+^{-s}(e^{-x^2/\sigma^2}) \qquad (8.1.2c)$$

$$\mathcal{G} = \|V(x)\|_{L^\infty} \qquad (8.1.2d)$$

$$L_{\text{NL}} = \left[\delta_{\text{NL}}^{-1}C_V(4\mathfrak{H}_+^s(\tilde{g}(\vec{x}))\mathfrak{H}_+^{-s}(e^{-x^2/\sigma^2}))\right]^{\frac{1}{1+\beta}} \qquad (8.1.2e)$$

$$k_{\text{sup,NL}}' = \frac{1}{2\alpha}\Gamma^{-1}\left(N, \frac{\delta_{\text{NL}}2^{N-7}\alpha^{2N-1}e^{-2\alpha(K-k_0)}}{(C_V')^2 N(K+k_0)(\mathfrak{H}_+^s(\tilde{g}(\vec{x}))\mathfrak{H}_+^{-s}(e^{-x^2/\sigma^2}))^2}\right) \qquad (8.1.2f)$$

$$L_F \geq (4C_V/3)^{1/(1+\beta)}T_{\mathfrak{M}}^{1/(1+\beta)}\delta_F^{-1/(1+\beta)} \qquad (8.1.2g)$$

$$\mathfrak{W}^+(\vec{b}, \delta_F, t) = \sigma\sqrt{1+t^2/\sigma^2}$$

$$\times \max\left\{\left(3\ln\left[\delta_F^{-1}\mathfrak{W}(L_{\text{in}}, k_{\text{sup}}, x_0, k_0, \sigma)(k_{\text{sup}}+1)T_{\mathfrak{M}}^{1/2}\right]\right)^{1/2}, \sqrt{2}\right\} \qquad (8.1.2h)$$

(with $\mathfrak{W}(L_{\text{in}}, k_{\text{sup}}, x_0, k_0, \sigma)$ defined below in (8.6.21b)). At this time we do not know how to validate Assumption 6, i.e. we can not control $k_{\text{inf}}(\delta_{\text{inf}})$.

We now turn to proving (8.1.2).

8.2 Assumption 1

This follows trivially from standard functional analysis. The operator $H = -(1/2)\Delta + V(x)$ is self adjoint and bounded below, so e^{iHt} is an isometric semigroup on L^2. Thus, the solution exists and is unique. This also implies that $\mathcal{L}(t) = 1$.

8.3 Assumption 2

This assumption holds due to conservation of energy, which allows us to prove that $\|\Psi(\vec{x}, t)\|_{H^1}$ is bounded.

Lemma 8.3.1 *We have the following bound on* $\|\Psi(\vec{x}, t)\|_{H^1}$:

$$\|\Psi(\vec{x}, t)\|_{H^1} \leq \sqrt{2}\sqrt{|E_0| + \|\Psi(x,0)\|_{L^2}^2 (\|V(x)\|_{L^\infty} + 1/2)} \qquad (8.3.1)$$

Proof Since $V(x)$ is real valued, (1.1.1) becomes a Hamiltonian system. Thus, $\langle \Psi(\vec{x}, t) | H\Psi(\vec{x}, t) \rangle$ is a conserved quantity. Therefore:

$$\langle \Psi(\vec{x}, t) | - (1/2)\Delta\Psi(\vec{x}, t) \rangle = \langle \Psi(x, 0) | H\Psi(x, 0) \rangle - \langle \Psi(\vec{x}, t) | V(x)\Psi(\vec{x}, t) \rangle$$

Multiplying by 2, adding $\|\Psi(x, 0)\|_{L^2}^2$ to both sides and then taking absolute values yields:

$$\|\Psi(\vec{x}, t)\|_{H^1}^2 \leq 2 \langle \Psi(x, 0) | H\Psi(x, 0) \rangle + 2 \|\Psi(\vec{x}, t)\|_{L^2}^2 \|V(x)\|_{L^\infty}^2 + \|\Psi(\vec{x}, t)\|_{L^2}$$

Applying conservation of mass to the $\Psi(\vec{x}, t)$ terms on the right, and then taking square roots yields the result we seek.   □

We note that $\left\| [1 - P^0_{[-K,K]^N;0}(k)]f(x) \right\|_{L^2} \leq \langle K \rangle^{-1} \|f(x)\|_{H^1}$. Combining this with Proposition 7.2.1 on p. 75, we have verified Assumption 2. Thus:

$$\left\| [1 - P^0_{[-K,K]^N;0}(k)]\Psi(\vec{x}, t) \right\|_{L^2}$$

$$\leq K^{-1}\sqrt{2}\sqrt{|E_0| + \|\Psi(x, 0)\|_{L^2}^2 (\|V(x)\|_{L^\infty} + 1/2)} \qquad (8.3.2)$$

Now, given δ_{sup}, we let

$$k_{\text{sup}} - \mathfrak{K}^s(k_{\text{sup}})$$

$$= \delta_{\text{sup}}^{-1} 2^{3/2} \mathfrak{H}_+^s(\tilde{g}(\vec{x})) \mathfrak{H}_+^{-s}(e^{-x^2/\sigma^2})\sqrt{|E_0| + \|\Psi(x, 0)\|_{L^2}^2 (\|V(x)\|_{L^\infty} + 1/2)}$$

Substituting this definition of k_{sup} into (8.3.2), we find that Proposition 7.2.1 is satisfied and therefore Assumption 2 is also satisfied.

One can, of course, use energy estimates (based on the fact that $\langle \Psi(\vec{x}, t) | H^s \Psi(\vec{x}, t) \rangle$ is conserved) in higher order Sobolev spaces to bound k_{sup} as well. In general, one can show that $k_{\text{sup}} \sim \delta_{\text{sup}}^{s-t}$ where s is the sobolev space in which we measure the error, and $t > s$ is some higher Sobolev space. However, the constants are difficult to control, due to the need to estimate many commutators, e.g. $[((1/2)\Delta)^a, V(x)^b]$ and higher orders.

8.4   Assumption 3

Since $V(x)$ is a bounded linear operator, we find that $\mathcal{G} = \|V(x)\|_{\mathcal{L}(H^s, H^s)}$. But $\|V(x)\|_{\mathcal{L}(H^s, H^s)}$ is given merely by $\|V(x)\|_{\mathcal{L}(H^s, H^s)} = \|V(x)\|_{W^{s,\infty}}$. In the case treated here ($s = 0$), we find simply that $\|V(x)\|_{\mathcal{L}(L^2, L^2)} = \|V(x)\|_{L^\infty}$.

8.5   Assumption 4

This follows from Assumption 2 combined with the fact that the potential is smooth and decays rapidly in space. We use the alternative assumption to Assumption 4 found on p. 76. We need to verify (7.2.2a) and (7.2.2b).

Bounds in Momentum
To verify (7.2.2a), we need to compute a bound on:

$$\left\| (1 - P^s_{[-M,M]^N;k_0}(\vec{x})) V(x) \Psi(\vec{x}, t) \right\|_{L^2}$$

We do this by using the fact that $\hat{V}(k)$ decays rapidly, combined with (8.3.2). We write:

$$\left\| (1 - P^0_{[-M,M]^N;k_0}(\vec{x})) V(x) \Psi(\vec{x}, t) \right\|_{L^2}$$

$$\leq \left\| (1 - P^0_{[-M,M]^N;k_0}(\vec{x})) [\hat{V}(k) \star P^0_{[-K,K]^N;k_0}(\vec{x}) \hat{\Psi}(k, t)] \right\|_{L^2}$$

$$+ \left\| (1 - P^0_{[-M,M]^N;k_0}(\vec{x})) [\hat{V}(k) \star (1 - P^0_{[-K,K]^N;k_0}(\vec{x})) \hat{\Psi}(k, t)] \right\|_{L^2}$$

$$\leq \left\| (1 - P^0_{[-M,M]^N;k_0}(\vec{x})) [\hat{V}(k) \star P^0_{[-K,K]^N;k_0}(\vec{x}) \hat{\Psi}(k, t)] \right\|_{L^2}$$

$$+ \| V(x) \|_{L^\infty} \left\| (1 - P^0_{[-K,K]^N;k_0}(\vec{x})) \hat{\Psi}(k, t) \right\|_{L^2}$$

$$\leq \left\| (1 - P^0_{[-M,M]^N;k_0}(\vec{x})) [\hat{V}(k) \star P^0_{[-K,K]^N;k_0}(\vec{x}) \hat{\Psi}(k, t)] \right\|_{L^2}$$

$$+ \| V(x) \|_{L^\infty} K^{-1} \sqrt{2} \sqrt{|E_0| + \| \Psi(x, 0) \|^2_{L^2} (\| V(x) \|_{L^\infty} + 1/2)} \qquad (8.5.1)$$

The last term can be made as small as necessary by making K large, which we will do shortly. We can calculate this (recalling (8.1.1b)) by:

$$\left\| (1 - P^0_{[-M,M]^N;k_0}(\vec{x})) [\hat{V}(k) \star P^0_{[-K,K]^N;k_0}(\vec{x}) \hat{\Psi}(k, t)] \right\|^2_{L^2}$$

$$\leq \int_{([-(M+k_0),(M+k_0)]^N)^C} \left| \int_{[-(K+k_0),(K+k_0)]^N} C'_V e^{-\alpha|\vec{k}-\vec{k}'|} \hat{\Psi}(\vec{k}', t) d\vec{k}' \right|^2 d\vec{k} \qquad (8.5.2)$$

The inner integral is the convolution of a compactly supported function with an exponentially decaying one. The result is exponentially decaying. The outer integral is then integrated over the tail of this exponentially decaying function, and is therefore exponentially small.

Lemma 8.5.1 *Suppose $|\vec{k}|_\infty \geq (K + k_0)$, in particular suppose that $|\vec{k}_j| \geq K + k_0$ with $j \in 1 \dots N$. We have the following bound on the inner integral:*

$$\left| \int_{[-(K+k_0),(K+k_0)]^N} C_V' e^{-\alpha |\vec{k}-\vec{k}'|} \hat{\Psi}(\vec{k}', t) d\vec{k}' \right|^2$$

$$\leq (C_V')^2 2(K + k_0) e^{-2\alpha(|\vec{k}|_\infty - K - k_0)} \|\Psi(x, 0)\|_{L^2}^2 \, \alpha^{-N+1}$$

Proof Since $|\vec{k}|_\infty \geq (K + k_0)$, there exists j so that $\left| \vec{k}_j \right| \geq K + k_0$. Suppose, without loss of generality, that $\vec{k}_j \geq K + k_0$ (the case when $\vec{k}_j \leq -K - k_0$ is just a change of coordinates). We can then calculate:

$$\left| \int_{[-(K+k_0),(K+k_0)]^N} (C_V')^2 e^{-\alpha |\vec{k}-\vec{k}'|} \hat{\Psi}(\vec{k}', t) d\vec{k}' \right|^2 \leq$$

$$\int_{[-(K+k_0),(K+k_0)]^N} (C_V')^2 e^{-2\alpha |\vec{k}-\vec{k}'|} \left\| \hat{\Psi}(\vec{k}, t) \right\|_{L^2}^2$$

$$\leq 2(K + k_0) e^{-2\alpha(\vec{k}_j - K - k_0)} \left\| \hat{\Psi}(\vec{k}, t) \right\|_{L^2}^2 \int_{[-(K+k_0),(K+k_0)]^{N-1}} e^{-2\alpha |\vec{k}-\vec{k}'|_1} d\vec{k}'$$

$$\leq 2(K + k_0) e^{-2\alpha(\vec{k}_j - K - k_0)} \left\| \hat{\Psi}(\vec{k}, t) \right\|_{L^2}^2 \int_{\mathbb{R}^{N-1}} e^{-2\alpha |\vec{k}-\vec{k}'|_1} d\vec{k}'$$

$$\leq 2(K + k_0) e^{-2\alpha(\vec{k}_j - K - k_0)} \left\| \hat{\Psi}(\vec{k}, t) \right\|_{L^2}^2 \alpha^{-N+1}$$

Finally, note that $\left\| \hat{\Psi}(\vec{k}, t) \right\|_{L^2} = \|\Psi(x, 0)\|_{L^2}$ and we are done. $\square$

Lemma 8.5.2 *The following equation holds.*

$$\int_{([-(M+k_0),(M+k_0)]^N)^C} e^{-2\alpha(|\vec{k}|_\infty - K - k_0)} d\vec{k}$$

$$= 2N(2\alpha)^{-N} e^{2\alpha(K-k_0)} \Gamma(N, 2\alpha(M + k_0))$$

$$\sim M^N e^{-2\alpha(M-K)} 2N(2\alpha)^{-N} \qquad (8.5.3)$$

In particular, (8.5.3) vanishes faster than $e^{-(2\alpha-\delta)M}$ for any $\delta > 0$.

Proof The set $\{\vec{k} : |\vec{k}|_\infty = u\}$ has surface area $2Nu^{N-1}$. Thus, we find that:

$$\int_{([-(M+k_0),(M+k_0)]^N)^C} e^{-2\alpha(|\vec{k}|_\infty - K - k_0)} d\vec{k}$$

$$= \int_{M+k_0}^{\infty} 2Nu^{N-1} e^{-2\alpha(u-K-k_0)} du = 2N e^{2\alpha(K-k_0)} \int_{M+k_0}^{\infty} u^{N-1} e^{-2\alpha u} du$$

$$= 2Ne^{2\alpha(K-k_0)} \int_{2\alpha(M+k_0)}^{\infty} (v/2\alpha)^{N-1} e^{-v} \frac{dv}{2\alpha}$$

$$= 2N(2\alpha)^{-N} e^{2\alpha(K-k_0)} \int_{2\alpha(M+k_0)}^{\infty} v^{N-1} e^{-v} dv$$

$$= 2N(2\alpha)^{-N} e^{2\alpha(K-k_0)} \Gamma(N, 2\alpha(M+k_0))$$

The asymptotics follow by applying (6.2.4) to $\Gamma(N, 2\alpha M)$.   □

We now apply Lemma 8.5.1 to the inner integral of (8.5.2), and Lemma 8.5.2 to the outer integral. We thus find that:

$$(8.5.2) \le (C_V')^2 2\alpha^{-N+1} 2N(2\alpha)^{-N}$$

$$\times \left\| \hat{\Psi}(\vec{k}, t) \right\|_{L^2}^2 (K+k_0) e^{2\alpha(K-k_0)} \Gamma(N, 2\alpha(M+k_0)) \qquad (8.5.4)$$

Thus:

$$(8.5.1) \le C_V' 2\alpha^{-(N+1)/2} N^{1/2} (2\alpha)^{-N/2}$$

$$\times \|\Psi(x, 0)\|_{L^2} \sqrt{K+k_0} e^{\alpha(K-k_0)} \sqrt{\Gamma(N, 2\alpha(M+k_0))}$$

$$+ \|V(x)\|_{L^\infty} K^{-1} \sqrt{2} \sqrt{|E_0| + \|\Psi(x, 0)\|_{L^2}^2 (\|V(x)\|_{L^\infty} + 1/2)} \qquad (8.5.5)$$

We will now make K and M sufficiently large.
We choose K in order to obtain:

$$\|V(x)\|_{L^\infty} K^{-1} \sqrt{2} \sqrt{|E_0| + \|\Psi(x, 0)\|_{L^2}^2 (\|V(x)\|_{L^\infty} + 1/2)}$$

$$\le \frac{1}{2} \frac{\delta_{NL} \|\Psi(\vec{x}, t)\|_{L^2}}{(4\mathfrak{H}_+^s(\tilde{g}(\vec{x}))\mathfrak{H}_+^{-s}(e^{-x^2/\sigma^2}))}$$

This yields:

$$K = 2^{3/2} (4\mathfrak{H}_+^s(\tilde{g}(\vec{x}))\mathfrak{H}_+^{-s}(e^{-x^2/\sigma^2}))$$

$$\times \frac{\|V(x)\|_{L^\infty} \sqrt{|E_0| + \|\Psi(x, 0)\|_{L^2}^2 (\|V(x)\|_{L^\infty} + 1/2)}}{\delta_{NL} \|\Psi(x, 0)\|_{L^2}}$$

We now select M so that:

$$C_V' 2\alpha^{-(N+1)/2} N^{1/2} (2\alpha)^{-N/2}$$

$$\times \|\Psi(x,0)\|_{L^2} \sqrt{K+k_0} e^{\alpha(K-k_0)} \sqrt{\Gamma(N, 2\alpha(M+k_0))}$$

$$\leq \frac{1}{2} \frac{\delta_{\mathrm{NL}} \|\Psi(\vec{x},t)\|_{L^2}}{(4\mathfrak{H}_+^s(\tilde{g}(\vec{x}))\mathfrak{H}_+^{-s}(e^{-x^2/\sigma^2}))}$$

This yields:

$$M = k_{\mathrm{sup,NL}}' = -k_0$$

$$+ (2\alpha)^{-1} \Gamma^{-1}\left(N, \frac{\delta_{\mathrm{NL}} 2^{N-7} \alpha^{2N-1} e^{-2\alpha(K-k_0)}}{(C_V')^2 N(K+k_0)(\mathfrak{H}_+^s(\tilde{g}(\vec{x}))\mathfrak{H}_+^{-s}(e^{-x^2/\sigma^2}))^2}\right) \qquad (8.5.6)$$

In terms of asymptotics, we find that $K = O(\delta_{\mathrm{NL}}^{-1})$ and $M = \Gamma^{-1}(N, e^{-2\alpha K}/K)$ $= \Gamma^{-1}(N, e^{-2\alpha \delta_{\mathrm{NL}}^{-1}} \delta_{\mathrm{NL}})$, thus $M = k_{\mathrm{sup,NL}}'$ grows at most $\delta_{\mathrm{NL}}^{-1-\delta}$ for any $\delta > 0$. Thus we have satisfied (7.2.2a).

The reader is hopefully convinced that one could (if sufficiently dedicated) verify these assumptions for stronger Sobolev spaces by controlling k_{sup} in these spaces. As remarked earlier, this is merely a matter of controlling various commuators, a tedius but doable process.

Bounds in Space

To verify (7.2.2b), we need to compute a bound on

$$\left\| P_{[-L_{\mathrm{NL}}, L_{\mathrm{NL}}]^N; x_0}^s (\vec{x}) \Psi(\vec{x}, t) \right\|_{L^2},$$

where we are free to choose L_{NL}'.

We can bound this by $\left\| (1 - P_{[-L_{\mathrm{NL}}, L_{\mathrm{NL}}]^N; x_0}^s (\vec{x})) V(x) \right\|_{L^\infty} \|\Psi(\vec{x}, t)\|_{L^2}$. Observe that (by (8.1.1a)):

$$\left\| (1 - P_{[-L_{\mathrm{NL}}, L_{\mathrm{NL}}]^N; x_0}^s (\vec{x})) V(x) \right\|_{L^\infty} \leq$$

$$\left\| 1 - P_{[-L_{\mathrm{NL}}, L_{\mathrm{NL}}]^N; x_0}^s (\vec{x}) \right\|_{L^\infty} C_V \langle L_{\mathrm{NL}}' \rangle^{-1-\beta} \leq C_V \langle L_{\mathrm{NL}}' \rangle^{-1-\beta}$$

Therefore, we find that in order to make

$$\left\| P_{[-L_{\mathrm{NL}}, L_{\mathrm{NL}}]^N; x_0}^s (\vec{x}) \Psi(\vec{x}, t) \right\|_{L^2} \leq \frac{\delta_{\mathrm{NL}} \|\Psi(\vec{x}, t)\|_{L^2}}{(4\mathfrak{H}_+^s(\tilde{g}(\vec{x}))\mathfrak{H}_+^{-s}(e^{-x^2/\sigma^2}))},$$

we need only let

$$L_{NL} = \left[\delta_{NL}^{-1} C_V (4\mathfrak{H}_+^s (\tilde{g}(\vec{x})) \mathfrak{H}_+^{-s} (e^{-x^2/\sigma^2}))\right]^{\frac{1}{1+\beta}}$$

Asymptotically, $L_{NL} = O((1/\delta_{NL})^{\frac{1}{1+\beta}})$.

If $V(x)$ decays exponentially, one can prove a similar estimate in which L_{NL} will behave like $O(\ln(1/\delta_{NL}))$.

8.6 Assumption 5

First, observe that $\mathcal{L}_{ext}(t) = 0$, since the Hamiltonian is linear.

We therefore need to verify (7.1.5a). The basic idea of why Assumption 5 is simply that the full flow differs from th free flow only near the potential. We now prove a lemma that makes this rigorous for compactly supported potentials.

Lemma 8.6.1 *Suppose* $V(x)$ *is real valued and compactly supported on* $B = [-L_{in}, L_{in}]^N$, *and suppose* $\Psi_0(x)$ *is smooth. Then:*

$$\left\| e^{iHt} \Psi_0(x) - e^{i(1/2)\Delta t} \Psi_0(x) \right\|_{L^2(B)}^2$$

$$\leq 4 \left(\int_0^t \int_{\partial B} [Flux\ Into\ B\ ]dt + \|\Psi_0(x)\|_{L^2(B)}^2 \right) \qquad (8.6.1)$$

The term [Flux Into B] is calculated under the free flow, i.e

$$[Flux\ Into\ B\ ] = \begin{cases} F(x, t), & F(x, t) > 0 \\ 0, & F(x, t) < 0 \end{cases}$$

with $F(x, t) = \Re[e^{i(1/2)\Delta t} \Psi_0(x)] \partial_n [e^{i(1/2)\Delta t} \Psi_0(x)]$ *and* ∂_n *the derivative normal to* ∂B. *Note that this refers to the flux into* B *only—the flux out of* B *is discarded in this calculation.*

Proof The basic idea of the proof is that $e^{i(1/2)\Delta t} \Psi_0(x)$ behaves like $e^{iHt} \Psi_0(x)$ except for those pieces of the wave which "touched" the potential $V(x)$. We use exact Dirichlet-to-Neumann boundary conditions, combined with a "Transfer Operator" to handle waves entering the box, following similar lines as in [36].

Following [36], we will decompose $\Psi(x, t)$ into 3 wave fields; waves which have not yet entered B, waves inside B and waves which have left B. Toward that end, let D be the exact Dirichlet-to-Neumann operator on ∂B. Also, define the positive flux operator F^+ by:

$$F^+\Psi(x,t) = \begin{cases} \Psi(x,t), & \Re\Psi(x,t)\partial_n\Psi(x,t) < 0, \\ 0, & \Re\Psi(x,t)\partial_n\Psi(x,t) > 0 \end{cases}, x \in \partial B \qquad (8.6.2)$$

and F^- similarly. Here, ∂_n is the normal derivative to ∂B oriented outward. Thus, $F^+\Psi(x,t)$ is the flux moving into B, and $F^-\Psi(x,t)$ is the flux moving out of B.

Let $\Psi_{0,e}$ solve the BVP (boundary value problem) on B^C:

$$i\partial_t\Psi_{0,e}(x,t) = -(1/2)\Delta\Psi_{0,e}(x,t) \qquad (8.6.3a)$$

$$\Psi_{0,e}(x,0) = \Psi_0(x) \qquad (8.6.3b)$$

$$\Psi_{0,e}(x,t) = F^+e^{i(1/2)\Delta t}\Psi_0(x), \; x \in \partial B \qquad (8.6.3c)$$

The boundary condition allows mass to enter B, but allows none to leave. The region B can be thought of as a perfectly absorbing sink for $\Psi_{0,e}(x,t)$. $\Psi_{0,e}(x,t)$ is the set of waves which have never entered B.

Let $\Psi_{0,i}(x,t)$ solve the BVP on B:

$$i\partial_t\Psi_{0,i}(x,t) = [-(1/2)\Delta + V(x)]\Psi_{0,i}(x,t) \qquad (8.6.4a)$$

$$\Psi_{0,i}(x,t) = \Psi_0(x) \qquad (8.6.4b)$$

$$\Psi_{0,i}(x,t) = D\Psi_{0,i}(x,t) + F^+e^{i(1/2)\Delta t}\Psi_0(x), \; x \in \partial B \qquad (8.6.4c)$$

The function $\Psi_{0,i}(x,t)$ represents the waves which have entered B, or were inside B to begin with. The Dirichlet-to-Neumann boundary allows waves to freely leave B.

Finally, let $\Psi_{1,e}(x,t)$ solve the BVP on B^C:

$$i\partial_t\Psi_{1,e}(x,t) = -(1/2)\Delta\Psi_{1,e}(x,t) \qquad (8.6.5a)$$

$$\Psi_{1,e}(x,0) = 0 \qquad (8.6.5b)$$

$$\Psi_{1,e}(x,t) = D\Psi_{0,i}(x,t), \quad x \in \partial B \qquad (8.6.5c)$$

The function $\Psi_{1,e}(x,t)$ represents waves which have left B.

The relation between all of this and $\Psi(x,t)$ is simple:

$$\Psi(x,t) = \Psi_{0,e}(x,t) + \Psi_{1,e}(x,t), \; x \in B^C$$

$$\Psi(x,t) = \Psi_{0,i}(x,t), \quad x \in B$$

The initial conditions clearly add up to $\Psi(x, 0) = \Psi_0(x)$ (recalling that $\Psi_1(x, 0) = 0$). The boundary conditions add up as well; $F^+ e^{i(1/2)\Delta t}\Psi_0(x)$ consists of incoming waves, which can only come from outside B, while $D\Psi_{0,i}(x, t)$ consists of outgoing waves, which can only come from inside B.

Now, to compute the bound, we note the following. If set set $V(x) = 0$, this changes the value of $\Psi_{0,i}(x, t)$ and $\Psi_{1,e}(x, t)$, but not $\Psi_{0,e}(x, t)$. This implies that:

$$\left\| \Psi(x, t) - \Psi^{V=0}(x, t) \right\|_{L^2(B^C)} \leq \left\| \Psi_{1,e}(x, t) - \Psi_{1,e}^{V=0}(x, t) \right\|_{L^2(B^C)}$$

$$\leq \left\| \Psi_{1,e}(x, t) \right\|_{L^2(B^C)} + \left\| \Psi_{1,e}^{V=0}(x, t) \right\|_{L^2(B^C)}$$

$$\leq \left\| \Psi_{1,e}(x, t) \right\|_{L^2(B^C)} + \left\| \Psi_{1,e}^{V=0}(x, t) \right\|_{L^2(B^C)}$$

$$+ \left\| \Psi_{0,i}(x, t) \right\|_{L^2(B)} + \left\| \Psi_{0,i}^{V=0}(x, t) \right\|_{L^2(B)}$$

$$\leq 2 \left(\int_0^t \int_{\partial B} [\text{Flux Into B }]dt + \|\Psi_0(x)\|_{L^2(B)} \right)^{1/2}$$

$$(8.6.6)$$

The last inequality simply follows by the continuity equation for $|\Psi(x, t)|^2$.   □

Remark 8.6.2 This result could have been proved by surrounding B by a sphere S, and expanding $\Psi_0(x)$ into spherical plane waves outside S (using, e.g. Kirchoff's formula). The outgoing plane waves would propagate identically, regardless of what happens inside S, although the incoming plane waves would be scattered. In fact, our proof ultimately reverts to this idea; though the Dirichlet-to-Neumann boundary and flux inflow condition allows us to short-circuit this analysis. A similar idea was used in [36] to construct Dirichlet-to-Neumann boundaries for multiple disconnected spheres.

We now need to use this to show that Assumption 5 holds. The main technical point is to show that sums of outgoing framelets are still outgoing.

We will now show that for a compactly supported potential, that sums of outgoing framelets are outgoing. We first show that sums of outgoing framelets with fixed velocity $\vec{b}k_0$ are outgoing. We then sum over all relevant $\vec{b}$. Toward that end, let B be a fixed set, and we define $V_c(x) = V(x)$ for $\vec{x} \in B$, and $V_c(x) = 0$ for $\vec{x} \notin B$. Similarly, define $H_c = -\Delta + V_c(x)$.

Proposition 8.6.3 *Define* $F_{\vec{b}} = \{(\vec{a}, \vec{b}') \in F : \vec{b}' = \vec{b}\}$ *to be some set of framelet indices with velocity* $\vec{b}$. *Suppose further that* $F_{\vec{b}}$ *is such that* $d(F_{\vec{b}} + \vec{b}kst, B) \geq R\sigma\sqrt{1 + t^2/\sigma^2}$, *for some* $R \geq \sqrt{2}$. *Then:*

$$
\left\| \left[e^{iH_c t} - e^{i(1/2)\Delta t} \right] \sum_{(\vec{a},\vec{b}) \in F_{\vec{b}}} \Psi_{(\vec{a},\vec{b})} \phi_{(\vec{a},\vec{b})}(\vec{x}) \right\|^2_{L^2(B)}
$$

$$
\leq \frac{4N2^N L^{N-1} 3}{\pi^{N/2} \sigma^N \left(1 - e^{-x_0/\sqrt{2}\sigma} \right)^N} \left[1 + \left(\left| \vec{b}_j k_0 \right| + R \right) t \right] e^{-R^2/2} \left(\sum_{\vec{a} \in \mathbb{Z}^N} \left| \Psi_{(\vec{a},\vec{b})} \right|^2 \right)
$$

$$
\tag{8.6.7}
$$

Here $d(x, Y)$ *denotes the distance between two sets.*

Proof By applying Lemma 8.6.1, it suffices to bound the flux into B under the free flow and the mass inside B at $t = 0$.

Preliminary Bounds
Make the definitions (for the purpose of this proof only):

$$
\alpha_{\vec{a}}^{0,1}(x, t) = \left(\left| \vec{b}_j k_0 \right| + \frac{|\vec{x} - \vec{b}k_0 t - \vec{a}x_0|_2}{\sigma\sqrt{1 + t^2/\sigma^4}} \right)^{0,1} \exp\left(-\frac{|\vec{x} - \vec{b}k_0 t - \vec{a}x_0|_2^2}{4\sigma^2(1 + t^2/\sigma^4)} \right)
$$

$$
\beta_{\vec{a}}(x, t) = \pi^{-N/4} \sigma^{-N/2} \frac{1}{(1 + t^2/\sigma^4)^{N/4}} \exp\left(-\frac{|\vec{x} - \vec{b}k_0 t - \vec{a}x_0|_2^2}{4\sigma^2(1 + t^2/\sigma^4)} \right)
$$

With these definitions, we find (by Cauchy-Schwartz):

$$
\left| \sum_{\vec{a} \in F_{\vec{b}}} \Psi_{(\vec{a},\vec{b})} \partial_n^{0,1} e^{i(1/2)\Delta t} \phi_{(\vec{a},\vec{b})}(\vec{x}) \right| \leq \left(\sup_{\vec{a} \in F_{\vec{b}}} \alpha_{\vec{a}}^{0,1}(x, t) \right) \left| \sum_{\vec{a} \in F_{\vec{b}}} \Psi_{(\vec{a},\vec{b})} \beta_{\vec{a}}(x, t) \right|
$$

$$
\leq \left(\sup_{\vec{a} \in F_{\vec{b}}} \alpha_{\vec{a}}^{0,1}(x, t) \right) \left(\sum_{\vec{a} \in F_{\vec{b}}} \left| \Psi_{(\vec{a},\vec{b})} \right|^2 \right)^{1/2} \left(\sum_{\vec{a} \in F_{\vec{b}}} \beta_{\vec{a}}(x, t) \right)^{1/2}
$$

$$
\leq \left(\sup_{\vec{a} \in F_{\vec{b}}} \alpha_{\vec{a}}^{0,1}(x, t) \right) \left(\sum_{\vec{a} \in \mathbb{Z}^N} \left| \Psi_{(\vec{a},\vec{b})} \right|^2 \right)^{1/2} \left(\sum_{\vec{a} \in \mathbb{Z}^N} \beta_{\vec{a}}(x, t)^2 \right)^{1/2} \tag{8.6.8}
$$

Note that for bounds on the initial condition, we use (8.6.8) with $\alpha_{\vec{a}}^0(x, t)$; the case with $\alpha_{\vec{a}}^1(x, t)$ will be used shortly to bound the flux. Here, ∂_n is the derivative normal to the box.

We can bound the sum over the $\beta_{\vec{a}}(x, t)$'s as follows:

$$\sum_{\vec{a} \in \mathbb{Z}^N} \beta_{\vec{a}}^2 \leq 3\pi^{-N/2}\sigma^{-N} \left(1 - e^{-x_0/\sqrt{2}\sigma}\right)^{-N}, \tag{8.6.9}$$

To prove this, first observe that $e^{-z^2} < (3/2)e^{-z}$. Thus:

$$\sum_{\vec{a} \in \mathbb{Z}^N} \beta_{\vec{a}}^2 = \sum_{\vec{a} \in \mathbb{Z}^N} \left(\frac{\pi^{-N/4}\sigma^{-N/2}}{(1 + t^2/\sigma^4)^{N/4}} \exp\left(-\frac{|\vec{x} - \vec{b}k_0 t - \vec{a}x_0|_2^2}{4\sigma^2(1 + t^2/\sigma^4)}\right)\right)^2$$

$$\leq \sum_{\vec{a} \in \mathbb{Z}^N} \frac{3\pi^{-N/2}\sigma^{-N}}{2(1 + t^2/\sigma^4)^{N/2}} \exp\left(-\frac{|\vec{x} - \vec{b}k_0 t - \vec{a}x_0|_2}{\sqrt{2}\sigma\sqrt{1 + t^2/\sigma^4}}\right)$$

$$\leq \frac{3\pi^{-N/2}\sigma^{-N}}{2(1 + t^2/\sigma^4)^{N/2}} \sum_{\vec{a} \in \mathbb{Z}^N} \exp\left(-\frac{|\vec{x} - \vec{b}k_0 t - \vec{a}x_0|_1}{\sqrt{2}\sigma\sqrt{1 + t^2/\sigma^4}}\right) \tag{8.6.10}$$

This sum is maximized when $\vec{x} - \vec{b}k_0 t = \vec{a}x_0$ for some $\vec{a}$. Thus:

$$(8.6.10) \leq \frac{3\pi^{-N/2}\sigma^{-N}}{2(1 + t^2/\sigma^4)^{N/2}} \sum_{\vec{a} \in \mathbb{Z}^N} \exp\left(-\frac{|\vec{a}x_0|_1}{\sqrt{2}\sigma\sqrt{1 + t^2/\sigma^4}}\right)$$

$$\leq \frac{3\pi^{-N/2}\sigma^{-N}}{2(1 + t^2/\sigma^4)^{N/2}} \left(\sum_{a \in \mathbb{Z}} \exp\left(-\frac{ax_0}{\sqrt{2}\sigma\sqrt{1 + t^2/\sigma^4}}\right)\right)^N$$

$$= \frac{3\pi^{-N/2}\sigma^{-N}}{(1 + t^2/\sigma^4)^{N/2}} \left(\frac{\exp\left(-\frac{x_0}{\sqrt{2}\sigma\sqrt{1+t^2/\sigma^4}}\right)}{1 - \exp\left(-\frac{x_0}{\sqrt{2}\sigma\sqrt{1+t^2/\sigma^4}}\right)}\right)^N$$

$$\leq \frac{3\pi^{-N/2}\sigma^{-N}}{(1 + t^2/\sigma^4)^{N/2}} \left(1 - \exp\left(-\frac{x_0}{\sqrt{2}\sigma\sqrt{1 + t^2/\sigma^4}}\right)\right)^{-N} \tag{8.6.11}$$

The time derivative of the last term is negative; thus:

$$(8.6.11) \leq 3\pi^{-N/2}\sigma^{-N} \left(1 - e^{-x_0/\sqrt{2}\sigma}\right)^{-N} \tag{8.6.12}$$

and (8.6.9) is proved.

Bounds on the Initial Mass

To bound the initial mass, square (8.6.8) to observe that at $t = 0$,

$$\left| \sum_{\vec{a} \in F_{\vec{b}}} \Psi_{(\vec{a}, \vec{b})} \phi_{(\vec{a}, \vec{b})}(\vec{x}) \right|^2 \leq \left(\sup_{\vec{a} \in F_{\vec{b}}} \alpha_{\vec{a}}^0(x, 0) \right)^2 \left(\sum_{\vec{a} \in \mathbb{Z}^N} |\Psi_{(\vec{a}, \vec{b})}|^2 \right) \left(\sum_{\vec{a} \in \mathbb{Z}^N} \beta_{\vec{a}}(x, t)^2 \right)$$

$$\leq \left(3\pi^{-N/2} \sigma^{-N} \left(1 - e^{-x_0/\sqrt{2}\sigma} \right)^{-N} \right)^2 \left(\sum_{\vec{a} \in \mathbb{Z}^N} |\Psi_{(\vec{a}, \vec{b})}|^2 \right) \left(\sup_{\vec{a} \in F_{\vec{b}}} \alpha_{\vec{a}}^0(x, 0) \right)^2$$

$$(8.6.13)$$

Since $d(F_{\vec{b}} + \vec{b}kst, B) \geq R\sigma\sqrt{1 + t^2/\sigma^2}$, we find that:

$$\left(\sup_{\vec{a} \in F_{\vec{b}}} \alpha_{\vec{a}}^0(x, 0) \right)^2 = \sup_{\vec{a} \in F_{\vec{b}}} [\alpha_{\vec{a}}^0(x, 0)]^2$$

$$= \sup_{\vec{a} \in F_{\vec{b}}} \exp \left(-\frac{|\vec{x} - \vec{b}k_0 t - \vec{a}x_0|_2^2}{2\sigma^2(1 + t^2/\sigma^4)} \right) \leq e^{-R^2/2} \qquad (8.6.14)$$

We now plug (8.6.14) into (8.6.13), and integrate over B to obtain:

$$\left\| \sum_{\vec{a} \in F_{\vec{b}}} \Psi_{(\vec{a}, \vec{b})} \phi_{(\vec{a}, \vec{b})}(\vec{x}) \right\|_{L^2(B)}^2$$

$$\leq \left(3\pi^{-N/2} \sigma^{-N} \left(1 - e^{-x_0/\sqrt{2}\sigma} \right)^{-N} \right)^2 \left(\sum_{\vec{a} \in \mathbb{Z}^N} |\Psi_{(\vec{a}, \vec{b})}|^2 \right) e^{-R^2/2} \qquad (8.6.15)$$

Bounds on the Flux

We will use bound the flux using (8.6.8). Since B is a box, we find that $\partial_n = \partial_{x_j}$ for some j (ignoring the corners, which have surface measure 0). Recalling (6.2.12) which provides an expression for $\partial_{x_j} e^{i(1/2)\Delta t} \phi_{(\vec{a}, \vec{b})}(\vec{x})$, we observe that:

$$\left| \partial_{x_j} e^{i(1/2)\Delta t} \phi_{(\vec{a}, \vec{b})}(\vec{x}) \right|$$

$$\leq \pi^{-N/4} \sigma^{-N/2} \left(\frac{|\vec{b}_j k_0|}{(1 + t^2/\sigma^4)^{N/4}} + \frac{|\vec{x}_j - \vec{b}_j k_0 t - \vec{a}_j x_0|}{\sigma(1 + t^2/\sigma^4)^{N/4 + 1/2}} \right)$$

$$\times \exp \left(-\frac{|\vec{x} - \vec{b}k_0 t - \vec{a}x_0|_2^2}{2\sigma^2(1 + t^2/\sigma^4)} \right) \qquad (8.6.16)$$

Recall (8.6.8); we will use it to bound the flux into B:

$$[\text{Flux Into B}](x) \leq \left| \Re \left(\sum_{(\vec{a},\vec{b}) \in F_{\vec{b}}} e^{i(1/2)\Delta t} \Psi_{(\vec{a},\vec{b})} \phi_{(\vec{a},\vec{b})}(\vec{x}) \right) \right.$$

$$\left. \times \, \partial_n^{0,1} \left(\sum_{(\vec{a},\vec{b}) \in F_{\vec{b}}} e^{i(1/2)\Delta t} \Psi_{(\vec{a},\vec{b})} \phi_{(\vec{a},\vec{b})}(\vec{x}) \right) \right|$$

$$\leq \left(\sup_{\vec{a} \in F_{\vec{b}}} \alpha_{\vec{a}}^0(x,t) \right) \left(\sup_{\vec{a} \in F_{\vec{b}}} \alpha_{\vec{a}}^1(x,t) \right) \left(\sum_{\vec{a} \in \mathbb{Z}^N} |\Psi_{(\vec{a},\vec{b})}|^2 \right) \left(\sum_{\vec{a} \in \mathbb{Z}^N} \beta_{\vec{a}}(x,t)^2 \right)$$

$$(8.6.17)$$

Recall (8.6.9), which bounds $\sum_{\vec{a} \in \mathbb{Z}^N} \beta_{\vec{a}}^2$. Note that if

$$\lambda = \frac{|\vec{x} - \vec{b}k_0 t - \vec{a}x_0|_2}{\sigma \sqrt{1 + t^2/\sigma^4}} \geq \sqrt{2}, \qquad (8.6.18)$$

then $\alpha_{\vec{a}}^{0,1}(x,t)$ is monotonic in the parameter λ. Thus, the $\vec{a}$ maximizing $\alpha_{\vec{a}}^0(x,t)$ is the same as the $\vec{a}$ maximizing $\alpha_{\vec{a}}^1(x,t)$. Therefore:

$$\left(\sup_{\vec{a} \in F_{\vec{b}}} \alpha_{\vec{a}}^0(x,t) \right) \left(\sup_{\vec{a} \in F_{\vec{b}}} \alpha_{\vec{a}}^1(x,t) \right) = \sup_{\vec{a} \in F_{\vec{b}}} \alpha_{\vec{a}}^0(x,t) \alpha_{\vec{a}}^1(x,t)$$

$$= \sup_{\vec{a} \in F_{\vec{b}}} \left(|\vec{b}_j k_0| + \frac{|\vec{x} - \vec{b}k_0 t - \vec{a}x_0|_2^2}{\sigma \sqrt{1 + t^2/\sigma^4}} \right) \exp \left(-\frac{|\vec{x} - \vec{b}k_0 t - \vec{a}x_0|_2^2}{2\sigma^2(1 + t^2/\sigma^4)} \right)$$

$$\leq \left(|\vec{b}_j k_0| + \frac{d(F_{\vec{b}} + \vec{b}k_0 t, B)}{\sigma \sqrt{1 + t^2/\sigma^4}} \right) \exp \left(-\frac{d(F_{\vec{b}} + \vec{b}k_0 t, B)^2}{2\sigma^2(1 + t^2/\sigma^4)} \right)$$

$$(8.6.19)$$

Since the function in the second line (8.6.19) is monotone decreasing in λ, we can maximize the function by choosing the smallest value of λ. This is given by $d(F_{\vec{b}} + \vec{b}k_0 t, B)/\sigma \sqrt{1 + t^2/\sigma^2}$, the distance between $F_{\vec{b}} + \vec{b}k_0 t$ and B, since we are only considering $x \in \partial B$. This is at least as large as R, provided $R \geq \sqrt{2}$ (since otherwise (8.6.18) might not be satisfied). Thus, we can bound the flux into B:

$$\int_{\partial B} [\text{Flux Into B}](x) dt$$

$$\leq \int_{\partial B} \left(\left| \vec{b}_j k_0 \right| + \frac{d(F_{\vec{b}} + \vec{b} k_0 t, B)}{\sigma \sqrt{1 + t^2/\sigma^4}} \right) \exp \left(-\frac{d(F_{\vec{b}} + \vec{b} k_0 t, B)^2}{2\sigma^2(1 + t^2/\sigma^4)} \right)$$

$$\times 3\pi^{-N/2} \sigma^{-N} \left(1 - e^{-x_0/\sqrt{2}\sigma} \right)^{-N} \left(\sum_{\vec{a} \in \mathbb{Z}^N} \left| \Psi_{(\vec{a},\vec{b})} \right|^2 \right) dx$$

$$\leq N 2^N L^{N-1} 3\pi^{-N/2} \sigma^{-N} \left(1 - e^{-x_0/\sqrt{2}\sigma} \right)^{-N}$$

$$\times \left(\sum_{\vec{a} \in \mathbb{Z}^N} \left| \Psi_{(\vec{a},\vec{b})} \right|^2 \right) \left(\left| \vec{b}_j k_0 \right| + \frac{d(F_{\vec{b}} + \vec{b} k_0 t, B)}{\sigma \sqrt{1 + t^2/\sigma^4}} \right) \exp \left(-\frac{d(F_{\vec{b}} + \vec{b} k_0 t, B)^2}{2\sigma^2(1 + t^2/\sigma^4)} \right)$$

$$\leq N 2^N L^{N-1} 3\pi^{-N/2} \sigma^{-N} \left(1 - e^{-x_0/\sqrt{2}\sigma} \right)^{-N} \left(\sum_{\vec{a} \in \mathbb{Z}^N} \left| \Psi_{(\vec{a},\vec{b})} \right|^2 \right)$$

$$\times \left(\left| \vec{b}_j k_0 \right| + R \right) \exp \left(-R^2/2 \right) \quad (8.6.20)$$

Integrating this result from 0 to t and multiplying by 4 yields (8.6.7).

$\square$

We can now add up the result of Proposition 8.6.3 over all relevant frequencies to obtain global bounds.

Proposition 8.6.4 *Let $F \subset \mathbb{Z}^N \times \mathbb{Z}^N$ be a set of framelets such that for any $(\vec{a}, \vec{b}) \in F$, $d(\vec{a} x_0 + \vec{b} k_0 t, B) \geq R\sigma \sqrt{1 + t^2/\sigma^2}$ with $R \geq \sqrt{2}$. Suppose further that if $\vec{b} \in F$, $|\vec{b} k_0|_\infty \leq k_{\text{sup}}$. Then:*

$$\left\| \left[e^{i H_c t} - e^{i(1/2)\Delta t} \right] \sum_{(\vec{a},\vec{b}) \in F} \Psi_{(\vec{a},\vec{b})} \phi_{(\vec{a},\vec{b})}(\vec{x}) \right\|_{L^2(B)}$$

$$\leq \mathfrak{V}(L_{\text{in}}, k_{\text{sup}}, x_0, k_0, \sigma) \|\Psi\|_{L^2_{\text{in}}} (k_{\text{sup}} + R) e^{-R^2/2} \langle t \rangle^{1/2} \quad (8.6.21a)$$

$$\mathfrak{V}(L_{\text{in}}, k_{\text{sup}}, x_0, k_0, \sigma) = \frac{\sqrt{N} 2^{N/2+1} L_{\text{in}}^{(N-1/2)} 3^{1/2} (2k_{\text{sup}}/k_0)^{N/2} \mathcal{B}_F}{\pi^{N/4} \sigma^{N/2} \left(1 - e^{-x_0/\sqrt{2}\sigma} \right)^{N/2}} \quad (8.6.21b)$$

Proof Define $F_{\vec{b}} = \{\vec{a} : (\vec{a}, \vec{b}) \in F\}$. Then:

$$\left\| \left[e^{iHt} - e^{i(1/2)\Delta t} \right] \sum_{(\vec{a},\vec{b}) \in F} \Psi_{(\vec{a},\vec{b})} \phi_{(\vec{a},\vec{b})}(\vec{x}) \right\|_{L^2_{\text{in}}(B)}$$

$$\leq \sum_{|\vec{b}k_0|_\infty \leq k_{\text{sup}}} \left\| \left[e^{iHt} - e^{i(1/2)\Delta t} \right] \sum_{(\vec{a},\vec{b}) \in F_{\vec{b}}} \Psi_{(\vec{a},\vec{b})} \phi_{(\vec{a},\vec{b})}(\vec{x}) \right\|_{L^2_{\text{in}}(B)}$$

$$\leq \sum_{\vec{b}} \left(\frac{4N2^N L_{\text{in}}^{N-1} 3}{\pi^{N/2} \sigma^N \left(1 - e^{-x_0/\sqrt{2}\sigma} \right)^N} \left(\sum_{\vec{a} \in \mathbb{Z}^N} \left| \Psi_{(\vec{a},\vec{b})} \right|^2 \right) \right.$$

$$\times \left[1 + \left(\left| \vec{b}_j k_0 \right| + R \right) t \right] e^{-R^2/2} \right)^{1/2}$$

$$\leq \frac{\sqrt{N} 2^{N/2+1} L_{\text{in}}^{(N-1/2)} 3^{1/2}}{\pi^{N/4} \sigma^{N/2} \left(1 - e^{-x_0/\sqrt{2}\sigma} \right)^{N/2}}$$

$$\times t^{1/2} \sum_{\vec{b}} \left(\sum_{\vec{a} \in \mathbb{Z}^N} \left| \Psi_{(\vec{a},\vec{b})} \right|^2 \right)^{1/2} \left[\left[1 + \left(\left| \vec{b}_j k_0 \right| + R \right) t \right] e^{-R^2/2} \right]^{1/2}$$

$$\leq \frac{\sqrt{N} 2^{N/2+1} L_{\text{in}}^{(N-1/2)} 3^{1/2}}{\pi^{N/4} \sigma^{N/2} \left(1 - e^{-x_0/\sqrt{2}\sigma} \right)^{N/2}} \langle t \rangle^{1/2}$$

$$\times \left(\sum_{(\vec{a},\vec{b}) \in F} \left| \Psi_{(\vec{a},\vec{b})} \right|^2 \right)^{1/2} \left(\sum_{\vec{b}} \left[\left(\left| \vec{b}_j k_0 \right| + R \right) e^{-R^2/2} \right] \right)^{1/2} \qquad (8.6.22)$$

The last line follows by Cauchy-Schwartz. The term $\left(\sum_{(\vec{a},\vec{b}) \in F} \left| \Psi_{(\vec{a},\vec{b})} \right|^2 \right)^{1/2}$ is bounded by $\mathcal{B}_F \|\Psi\|_{L^2}$. The latter term satisfies the bounds:

$$\left(\sum_{|\vec{b}k_0|_\infty \leq k_{\text{sup}}} \left[\left(|\vec{b}k_0|_\infty + R \right) e^{-R^2/2} \right] \right) \leq (2k_{\text{sup}}/k_0)^N (k_{\text{sup}} + R) e^{-R^2/2} \qquad (8.6.23)$$

Substituting (8.6.23) into (8.6.22) and simplifying yields the result we seek.   □

We must now show that $e^{iH_c t} \approx e^{iH t}$. This is done as follows:

Proposition 8.6.5 *We have the following bound:*

$$\left\| e^{iH t} - e^{iH_c t} \right\|_{\mathcal{L}(L^2, L^2)} \leq \exp\left(\|V(x) - V_c(x)\|_{L^\infty} t \right) - 1 \tag{8.6.24}$$

Proof By Duhamel's principle, we have:

$$e^{iH_c t} \Psi_0(x) = e^{iH t} \Psi_0 + i \int_0^t e^{iH(t-s)} (V(x) - V_c(x)) e^{iH_c s} \Psi_0 ds$$

Rearranging implies that:

$$e^{iH_c t} \Psi_0(x) - e^{iH t} \Psi_0 = i \int_0^t e^{iH(t-s)} (V(x) - V_c(x)) e^{iH s} \Psi_0 ds$$

$$+ i \int_0^t e^{iH(t-s)} (V(x) - V_c(x)) \left[e^{iH_c s} \Psi_0 - e^{iH s} \Psi_0 \right] ds$$

Noting that $e^{iH t}$ and $e^{iH_c t}$ are unitary, and setting $f(t) = \left\| e^{iH_c t} \Psi_0(x) - e^{iH t} \Psi_0 \right\|_{L^2}$, we find that:

$$f(t) \leq t \|V(x) - V_c(x)\|_{L^\infty} + \|V(x) - V_c(x)\|_{L^\infty} \int_0^t f(s) ds$$

Lemma 9.1.2 (Gronwall's lemma) yields (8.6.24). □

To verify Assumption 5, we must now find $L_F, \mathfrak{W}^+(\vec{b}, \delta_F, t)$ sufficiently large to ensure that (7.1.5a) holds.

Toward that end, note that:

$$\left\| e^{iH t} f(x) - e^{i(1/2)\Delta t} f(x) \right\|_{L^2}$$

$$\leq \left\| e^{iH t} f(x) - e^{iH_c t} f(x) \right\|_{L^2} + \left\| e^{iH_c t} f(x) - e^{i(1/2)\Delta t} f(x) \right\|_{L^2} \tag{8.6.25}$$

To bound the first term by $\delta_F/2$, we simply use Proposition 8.6.5 and choose L_F sufficiently large so that $\exp\left(\|V(x) - V_c(x)\|_{L^\infty} T_{\mathfrak{M}} \right) - 1 \leq \delta/2$, or

$$\|V(x) - V_c(x)\|_{L^\infty} \leq T_{\mathfrak{M}}^{-1} \ln(1 + \delta_F/2).$$

Using (8.1.1a) to bound $\|V(x) - V_c(x)\|_{L^\infty}$, we find that this can be accomplished if:

$$L_F \geq \left(\frac{T_{\mathfrak{M}} C_V}{\ln(1 + \delta_F/2)} \right)^{1/(1+\beta)} \geq (4C_V/3)^{1/(1+\beta)} T_{\mathfrak{M}}^{1/(1+\beta)} \delta_F^{-1/(1+\beta)} \tag{8.6.26}$$

and $\delta_F/2 < 1$. To obtain this we used the fact that $\ln(1 + x) \geq 3x/4$ for $0 < x < 1$ to bound $\ln(1 + \delta_F/2)$ from below.

We must now bound $\left\|e^{iH_ct} f(x) - e^{i(1/2)\Delta t} f(x)\right\|_{L^2}$. To do this, we apply Proposition 8.6.4, using the fact that $f(x) = \Psi_{(\vec{a},\vec{b})}\phi_{(\vec{a},\vec{b})}(\vec{x})$,

Note that $re^{-r^2/2} \leq 3e^{-r^2/3}$. We find that (by (8.6.21)):

$$\left\|\left[e^{iHt} - e^{i(1/2)\Delta t}\right] \sum_{(\vec{a},\vec{b})\in F} \Psi_{(\vec{a},\vec{b})}\phi_{(\vec{a},\vec{b})}(\vec{x})\right\|_{L^2(B)}$$

$$\leq \mathfrak{V}(L_{in}, k_{sup}, x_0, k_0, \sigma)\, \|\Psi\|_{L^2_{in}}\, (k_{sup} + 1)e^{-R^2/3}\langle t\rangle^{1/2} \qquad (8.6.27)$$

We therefore choose

$$\mathfrak{W}^+(\vec{b}, \delta_F, t)$$

$$= \left(3 \ln\left[\delta_F^{-1} 2\mathfrak{V}(L_{in}, k_{sup}, x_0, k_0, \sigma)(k_{sup} + 1)\langle T_{\mathfrak{M}}\rangle^{1/2}\right]\right)^{1/2} \sigma\sqrt{1 + t^2/\sigma^2}. \qquad (8.6.28)$$

or $\mathfrak{W}^+(\vec{b}, \delta_F, t) = \sqrt{2}\sigma\sqrt{1 + t^2/\sigma^2}$, whichever is larger. This implies that

$$\left\|e^{iH_ct} f(x) - e^{i(1/2)\Delta t} f(x)\right\|_{L^2} \leq \delta_F/2.$$

8.7   Assumption 6

This is the only assumption we do not know how to verify.

Assumption 6 has two major pieces which need to be verified. First, we must show that framelets outside the box decay away from the propagation set where $\vec{a}x_0 \parallel \vec{b}k_0$. Second, we must show that the mass of outgoing framelets with velocity $|\vec{b}k_0|_\infty$ below $2\sqrt{N}k_{inf}$ is small.

It is the second fact which we do not know how to prove, and indeed which we do not believe to be true in general (Sect. 7.6.4 sketches a possible counterexample). We are uncertain at this time how to show that the amount of mass below $2\sqrt{N}k_{inf}$ is small, and in some cases it is not small. This problem can be treated in the context of the TDPSF by a multiscale extension of this method, which is described in [63]. This extension reduces the computational complexity to $O(\log k_{inf})$ rather than $O(k_{inf}^{-1})$, and does not require apriori assumptions on what k_{inf} is.

We believe the first can be verified by using pseudoconformal-type estimates, which we will sketch out below.

We now sketch an argument suggesting that waves cluster on waves where $\vec{x} \parallel \vec{k}$. Recall that in Remark 7.2.4, we provided an argument suggesting that if

$\|(\vec{x} - it\nabla)f(\vec{x})\|_{L^2}$ was bounded, then the mass of $f(\vec{x})$ sitting on framelets with $|\vec{a}x_0 - \vec{b}k_0 t|_2 \gg 0$ is small.

We now suppose that $\Psi(x, t)$ is located strictly on positive energies, i.e. $\chi_{[k_{\inf}, \infty)}(H)\Psi(x, t) = \Psi(x, t)$. Let us also suppose that $\langle \vec{x}\rangle^2 V(\vec{x})$ decays rapidly. This suggests to us that $\left\||\vec{x}|_2^2 V(\vec{x})\Psi(x, t)\right\|_{L^2} \le \text{const}\, t^{-3/2}$.

We can then decompose $\Psi(x, t)$ by Duhamel in the following way:

$$\Psi(x, t) = e^{i(1/2)\Delta t}\Psi(x, 0) + \int_{jt/n}^{(j+1)t/n} e^{i(1/2)\Delta(t-t')}V(x)\Psi(x, t')dt'$$

We then observe that:

$$\left\|(x - it\nabla)e^{i(1/2)\Delta t}\Psi(x, 0)\right\|_{L^2} = \left\||\vec{x}|_2^2\Psi(x, 0)\right\|_{L^2}$$

In addition, we find that:

$$\left\|(x - i(t - t')\nabla)e^{i(1/2)\Delta(t-t')}V(x)\Psi(x, t')\right\|_{L^2}$$
$$= \left\|e^{i(1/2)\Delta(t-t')}|\vec{x}|_2^2 V(x)\Psi(x, t')\right\|_{L^2}$$
$$= \left\||\vec{x}|_2^2 V(x)\Psi(x, t')\right\|_{L^2} \le \text{const}(t')^{-3/2}$$

We then observe that this suggests that the framelet coefficients of $e^{i(1/2)\Delta(t-t')}V(x)\Psi(x, t')$ are also small when $\vec{a}x_0 \perp \vec{b}k_0$. This indicates that:

$$\left\|\mathcal{P}_{\vec{a}\perp\vec{b}}\Psi(x, t)\right\|_{L^2} \le \text{small}\left\||\vec{x}|_2^2\Psi(x, 0)\right\|_{L^2} + \text{small}\int_0^t \text{const}(t')^{-3/2}dt'$$

$$(8.7.1)$$

This argument, which we believe can be made rigorous, suggests why we believe that all of our assumptions can be verified for the case of linear, time independent potentials.

Chapter 9
Proof of Theorem 7.5.6

In this part, we construct a bound on the difference between the free propagator and the box propagator acting on waves which are not outgoing:

$$\|(\mathfrak{U}(t) - \mathfrak{U}_b(t))\mathcal{P}_{\text{NECC} \cap \text{BB}}\Psi_0(x)\|_{H^s}$$

We do this by Duhamel's principle and Gronwall, and use the fact that the nonlinearity is locally Lipschitz (Assumption 3). The bound on this term is summarized in Theorem 7.5.6 in the next section.

We first define three functions, $E(t)$, $Q(t)$ and $Q_b(t)$ which we will use to construct error bounds.

Definition 9.0.1 Let $\Psi(\vec{x}, t)$ be a solution to (1.1.1) on $\mathbb{R}^N$, and $\Psi_b(\vec{x}, t)$ be a solution to (1.1.1) on $[-L_{\text{in}}, L_{\text{in}}]^N$. Suppose that

$$\Psi(\vec{x}, 0) = \Psi_b(\vec{x}, 0) = \mathcal{P}_{\text{NECC} \cap \text{BB}}\Psi_0(x)$$

for some $\Psi_0(x)$.

We define the free error function to be some function $E(t)$ for which:

$$\left\|(e^{i(1/2)\Delta t} - e^{i(1/2)\Delta_b t})\Psi(x, 0)\right\|_{H_b^s} \leqslant E(t) \tag{9.0.1}$$

We define the interaction error to be functions $Q(t)$ (or $Q_b(t)$) for which:

$$\left\|\int_0^t (e^{i(1/2)\Delta(t-t')} - e^{i(1/2)\Delta_b(t-t')})g(t', \vec{x}, \Psi(x, t'))\Psi(x, t')dt'\right\|_{H_b^s} \leq Q(t) \tag{9.0.2a}$$

$$\left\|\int_0^t (e^{i(1/2)\Delta(t-t')} - e^{i(1/2)\Delta_b(t-t')})g(t', \vec{x}, \Psi_b(x, t'))\Psi_b(x, t')dt'\right\|_{H_b^s} \leq Q_b(t) \tag{9.0.2b}$$

© The Author(s), under exclusive license to Springer Nature Singapore Pte Ltd. 2023
A. Soffer et al., *Time Dependent Phase Space Filters*, SpringerBriefs on PDEs and Data Science, https://doi.org/10.1007/978-981-19-6818-1_9

We will write our estimates in terms of these functions. We show that (7.5.18) is consistent with Definition 9.0.1.

The rest of Sect. 9 is devoted to proving various pieces of Theorem 7.5.6. We prove (7.5.17) in Sect. 9.1. The estimate (7.5.18a) is done in in Sect. 9.2 (Proposition 9.2.5) while (7.5.18b) is proved in Sect. 9.4 (Proposition 9.4.2).

9.1   Estimates of $E(t)$, $Q(t)$

Here, we prove the estimates (7.5.17a) and (7.5.17b) assuming that $E(t)$ and $Q(t)$ are known.

We state the result in a more general manner, which we believe will also be useful for proving short time error bounds for other types of absorbing boundary conditions.

Theorem 9.1.1 *Let $\Psi_0(x) \in H^s$. Let $g(t, \vec{x}, \cdot)$ satisfy Assumption 3. Let $E(t)$ be defined by (9.0.1), and $Q(t)$, $Q_b(t)$ by (9.0.2). Then the following holds:*

$$\|(\mathfrak{U}(t) - \mathfrak{U}_b(t))\Psi_0(x)\|_{H_b^s} \leq (E(t) + Q(t)) + \mathcal{G}e^{\mathcal{G}t} \star (E(t) + Q(t)) \text{ (apriori)}$$

$$(9.1.1a)$$

$$\|(\mathfrak{U}(t) - \mathfrak{U}_b(t))\Psi_0(x)\|_{H_b^s} \leq (E(t) + Q_b(t)) + \mathcal{G}e^{\mathcal{G}t} \star (E(t) + Q_b(t)) \text{ (aposteriori)}$$

$$(9.1.1b)$$

Lemma 9.1.2 (Gronwall) *Let $y(t)$ satisfy the inequality:*

$$y(t) \leq p(t) + C \int_0^t y(t)dt \qquad (9.1.2)$$

$y(t)$ satisfies the bound:

$$y(t) \leq p(t) + Ce^{Ct} \int_0^t e^{-Cs} p(s)ds \qquad (9.1.3)$$

Proof of Theorem 9.1.1 We use Duhamel. We observe the following equality:

$$\Psi(t) - \Psi_b(t) = e^{i(1/2)\Delta t}\varphi(x) - e^{i(1/2)\Delta_b t}\varphi(x)$$

$$+ i \int_0^t [e^{i(1/2)\Delta(t-s)}g(s, \vec{x}, \Psi(s))\Psi(s) - e^{i(1/2)\Delta_b t - s}g(s, \vec{x}, \Psi_b(s))\Psi_b(s)]ds$$

We then add and subtract $e^{i(1/2)\Delta(t-s)}g(s, \vec{x}, \Psi_b(s))\Psi_b(s)$ under the integral sign, and take norms in H^s to obtain:

$$\|\Psi(t) - \Psi_b(t)\|_{H_b^s} \leqslant E_s(t)$$

$$+ \left\| \int_0^t [e^{i(1/2)\Delta(t-s)} - e^{i(1/2)\Delta_b t-s}]g(s, \vec{x}, \Psi_b(s))\Psi_b(s)ds \right\|_{H_b^s} +$$

$$\left\| \int_0^t e^{i(1/2)\Delta(t-s)}[g(s, \vec{x}, \Psi(s))\Psi(s) - g(s, \vec{x}, \Psi_b(s))\Psi_b(s)]ds \right\|_{H_b^s}$$

We then observe that

$$\|g(s, \vec{x}, \Psi(s))\Psi(s) - g(s, \vec{x}, \Psi_b(s))\Psi_b(s)\|_{H^s} \leq \mathcal{G} \|\Psi(s) - \Psi_b(s)\|_{H^s}$$

and also that the first term is $Q_b(t)$. Gronwall's Lemma 9.1.2 gives us (9.1.1b). Estimate (9.1.1a) follows in much the same way, except that we add and subtract $e^{i(1/2)\Delta_b t-s}g(s, \vec{x}, \Psi(s))\Psi(s)$ instead.   □

Proof of Lemma 9.1.2 In the case of equality, we have:

$$y(t) = p(t) + C \int_0^t y(t)dt$$

Laplace transformation yields:

$$Y(z) = P(z) - C\frac{Y(z)}{z}$$

Or equivalently:

$$Y(z) = \left(1 + \frac{C}{z + C}\right) P(z)$$

Inverting the Laplace transform and collecting residues yields the result we seek:

$$y(t) = e^{Ct} \int_0^t e^{-Cs}\frac{dp(s)}{ds}ds = p(t) + Ce^{Ct} \int_0^t e^{-Cs}p(s)ds$$

□

In the event that $g(t, \vec{x}, \Psi)\Psi = V(x, t)\Psi(x)$ a sharper estimate holds. This can not be shown to hold in the nonlinear case—indeed, counterexamples exist.

Theorem 9.1.3 *Let* $\Psi(x, t = 0) \in H^s$ *be an initial condition of* (1.1.1), *where* $g(t, \vec{x}, \Psi)\Psi = V(x, t)\Psi(x)$ *(that is, a "linear nonlinearity"). Suppose that the equation*

$$i\partial_t \Psi_b(\vec{x}, t) = (-(1/2)\Delta_b + V(\vec{x}, t)) \Psi_b(\vec{x}, t)$$

satisfies the energy conservation law $\|\Psi_b(x, t)\|_{H^s} \leqslant \alpha(t) \|\Psi_b(x, 0)\|_{H^s}$. *Then we find:*

$$\|\Psi(\vec{x}, t) - \Psi_b(\vec{x}, t)\|_{H_b^s}$$

$$\leq \alpha(t) \|\Psi(x, 0) - \Psi_b(x, 0)\|_{H_b^s} + \int_0^t \alpha(t - t') \|S(\vec{x}, t)\|_{H_b^s} dt' \qquad (9.1.4)$$

where:

$$S(x, t) = \left[(e^{i(1/2)\Delta_b t} - e^{i(1/2)\Delta t})\Psi(x, 0) \right.$$

$$\left. + i \int_0^t \left(e^{i(1/2)\Delta_b(t-t')} - e^{i(1/2)\Delta(t-t')} \right) V(x, t')\Psi(x, t')dt \right] \qquad (9.1.5a)$$

$$s(x, t) = i\partial_t S(x, t) \qquad (9.1.5b)$$

In particular, observe that $\|S(x, t)\|_{H^s} \leqslant E(t) + Q(t)$, *so to bound the error, it is sufficient to construct* $E(t)$ *and* $Q(t)$.

Proof We write $\Psi_b(x, t) = \Psi(x, t) + e(x, t)$ where $e(x, t)$ is the error. We then subtract the Duhamel equation for $\Psi_b(x, t)$ from the Duhamel equation for $\Psi(x, t)$ to obtain:

$$e(x, t) = e^{i(1/2)\Delta_b(t-t')}e(x, 0) + i \int_0^t e^{i(1/2)\Delta_b(t-t')} V(x, t')e(x, t')dt'$$

$$+ (e^{i(1/2)\Delta_b t} - e^{i(1/2)\Delta t})\Psi(x, 0)$$

$$+ i \int_0^t \left(e^{i(1/2)\Delta_b(t-t')} - e^{i(1/2)\Delta(t-t')} \right) V(x, t')\Psi(x, t')dt'$$

If we apply $i\partial_t$ to this equation, we observe that:

$$i\partial_t e(x, t) = (-(1/2)\Delta_b + V(x, t)) e(x, t) + S(x, t)$$

Taking norms and bringing them under the integral sign gives us the result we seek.

$\square$

9.2   Estimates of $E(t)$

Here, the bound (7.5.18a) on $E(t)$ is constructed from the framelet decomposition and the fact that $\Psi(x, 0)$ is given by framelets which are in NECC ∩ BB. We further split this up into framelets which are in BAD^C ∩ NECC ∩ BB and BAD ∩ NECC ∩ BB. We then add the results together to obtain the estimate.

Lemma 9.2.1 *Let $\{\phi_j\}$ be a frame with frame bounds $\mathcal{A}_F, \mathcal{B}_F$ and with per-framelet error bounds $\{\mathfrak{E}_j^s(t)\}$. Suppose J is a finite set of framelet indices. Then:*

$$\left\| (e^{i(1/2)\Delta t} - e^{i(1/2)\Delta_b t}) \sum_{j \in J} \Psi_j \phi_j(x) \right\|_{H^s} \leq \sum_{j \in J} \left| \Psi_j \right| \mathfrak{E}_j^s(t)$$

$$\leq \mathcal{A}_F^{-1} \sqrt{\sum_{j \in J} \left| \mathfrak{E}_j^s(t) \right|^2} \, \|\Psi\|_{L^2} \leq \mathcal{A}_F^{-1} \sqrt{|J|} \sup_{j \in J} \mathfrak{E}_j^s(t) \, \|\Psi\|_{L^2} \qquad (9.2.1)$$

Here, $|J|$ represents the cardinality of J. The same result holds if we replace $(e^{i(1/2)\Delta t} - e^{i(1/2)\Delta_b t})$ by $\chi_{[-L_{\text{in}}, L_{\text{in}}]^N} e^{i(1/2)\Delta t}$ and $\mathfrak{E}_j^s(t)$ by $\mathcal{R}_j^s(t)$.

Proof The triangle inequality yields:

$$\left\| (e^{i(1/2)\Delta t} - e^{i(1/2)\Delta_b t}) \sum_{j \in J} \Psi_j \phi_j(x) \right\|_{H^s} \leqslant \sum_{j \in J} \left| \Psi_j \right| \mathfrak{E}_j^s(t)$$

We have a sharp bound:

$$\leqslant \sum_{j \in J} \left| \Psi_j \right| \mathfrak{E}_j^s(t) \leq \sqrt{\sum_{j \in J} \left| \Psi_j \right|^2} \sqrt{\sum_{j \in J} \left| \mathfrak{E}_j^s(t) \right|^2} \leq \mathcal{A}_F^{-1} \|\Psi\|_{L^2} \sqrt{\sum_{j \in J} \left| \mathfrak{E}_j^s(t) \right|^2}$$

We obtain a suboptimal (although still reasonably useful) bound:

$$\sqrt{\sum_{j \in J} \left| \mathfrak{E}_j^s(t) \right|^2} \leq \sqrt{|J|} \sup_{j \in J} \mathfrak{E}_j^s(t)$$

This yields the result we seek. The proof with $\mathcal{R}_j^s(t)$ instead of $\mathfrak{E}_j^s(t)$ is identical, but with $e^{i(1/2)\Delta t}$ replacing $(e^{i(1/2)\Delta t} - e^{i(1/2)\Delta_b t})$. $\qquad \square$

Remark 9.2.2 For practical purposes, the estimate $\sqrt{\sum_{j\in J}\left|\mathfrak{E}_j^s(t)\right|^2}$ should be used rather than $A_F^{-1}\sqrt{|J|}\sup_{j\in J}\mathfrak{E}_j^s(t)\|\Psi\|_{L^2}$. For any given set of parameters it is simple to compute, and gives a precise estimate (which does not grow with L). The cruder estimate is included to demonstrate that the estimate is nontrivial.

We now apply Lemma 9.2.1 to obtain the following result dealing with framelets in $\text{NECC}\cap\text{BB}\cap\text{BAD}^C$.

Proposition 9.2.3 *Let $\Psi_0(x)$ satisfy Assumption 2. Then we find:*

$$\left\|(e^{i(1/2)\Delta t}-e^{i(1/2)\Delta_b t})\mathcal{P}_{\text{BAD}^C\cap\text{NECC}\cap\text{BB}}\Psi_0(x)\right\|_{H_b^s}\le\widehat{\mathcal{E}}(t)\,\|\Psi\|_{L^2}\qquad(9.2.2)$$

Proof Compute:

$$\left\|(e^{i(1/2)\Delta t}-e^{i(1/2)\Delta_b t})\sum_{(\vec{a},\vec{b})\in\text{BAD}^C\cap\text{NECC}\cap\text{BB}}\Psi_{0(\vec{a},\vec{b})}\phi_{(\vec{a},\vec{b})}(\vec{x})\right\|_{H_b^s}$$

$$\le\sum_{(\vec{a},\vec{b})\in\text{BAD}^C\cap\text{NECC}\cap\text{BB}}\left|\Psi_{0(\vec{a},\vec{b})}\right|\left\|(e^{i(1/2)\Delta t}-e^{i(1/2)\Delta_b t})\phi_{(\vec{a},\vec{b})}(\vec{x})\right\|_{H_b^s}$$

$$\le\sqrt{\sum_{(\vec{a},\vec{b})\in\text{BAD}^C\cap\text{NECC}\cap\text{BB}}\mathfrak{E}_{(\vec{a},\vec{b})}(t)^2}$$

$$\times\sqrt{\sum_{(\vec{a},\vec{b})\in\text{BAD}^C\cap\text{NECC}\cap\text{BB}}\left|\Psi_{0(\vec{a},\vec{b})}\right|^2}\le\widehat{\mathcal{E}}(t)\,\|\Psi_0\|_{L^2}$$

Here we used the fact that

$$\sqrt{\sum_{(\vec{a},\vec{b})\in\text{BAD}^C\cap\text{NECC}\cap\text{BB}}\left|\Psi_{0(\vec{a},\vec{b})}\right|^2}\le A_F^{-1}\,\|\Psi_0\|_{L^2}$$

and the definition of $\widehat{\mathcal{E}}(t)$ (Definition 7.5.1 on p. 80). $\qquad\square$

Proposition 9.2.4 *Let the parameters k_{inf}, w and T_{st} satisfy (7.5.16). Let $\Psi_0(x)$ satisfy Assumption 6. Then the following estimate holds:*

$$\left\|(e^{i(1/2)\Delta t}-e^{i(1/2)\Delta_b t})\mathcal{P}_{\text{BAD}\cap\text{NECC}\cap\text{BB}}\Psi_0(x)\right\|_{H_b^s}\le 2\delta_{\text{inf}}\qquad(9.2.3)$$

This result is slightly trickier, and Sect. 9.3 is devoted to the proof. We now arrive at the bound on $E(t)$:

Proposition 9.2.5 Let $\Psi_0(x)$ satisfy Assumption 6, and let L_{in}, T_{st} and w satisfy (7.5.16). Then:

$$\left\| (e^{i(1/2)\Delta t} - e^{i(1/2)\Delta_b t}) \mathcal{P}_{\text{BB} \cap \text{NECC}} \Psi_0(x) \right\|_{H_b^s}$$
$$\leq \widehat{\mathcal{E}}(t) \|\Psi_0(x)\|_{L^2} + 2\delta_{\text{inf}} = E(t) \qquad (9.2.4)$$

Proof Observe that

$$\mathcal{P}_{\text{BB} \cap \text{NECC}} \Psi_0(x) = \mathcal{P}_{\text{BAD} \cap \text{BB} \cap \text{NECC}} \Psi_0(x) + \mathcal{P}_{\text{BAD}^C \cap \text{BB} \cap \text{NECC}} \Psi_0(x) \qquad (9.2.5)$$

We therefore apply $(e^{i(1/2)\Delta t} - e^{i(1/2)\Delta_b t})$ to (9.2.5), then take the norm in H_b^s and use the triangle inequality, to obtain:

$$\left\| (e^{i(1/2)\Delta t} - e^{i(1/2)\Delta_b t}) \mathcal{P}_{\text{BB} \cap \text{NECC}} \Psi_0(x) \right\|_{H_b^s}$$
$$\leq \left\| (e^{i(1/2)\Delta t} - e^{i(1/2)\Delta_b t}) \mathcal{P}_{\text{BAD} \cap \text{BB} \cap \text{NECC}} \Psi_0(x) \right\|_{H_b^s}$$
$$+ \left\| (e^{i(1/2)\Delta t} - e^{i(1/2)\Delta_b t}) \mathcal{P}_{\text{BAD}^C \cap \text{BB} \cap \text{NECC}} \Psi_0(x) \right\|_{H_b^s} \qquad (9.2.6)$$

Then apply Proposition 9.2.3 to the last term and Proposition 9.2.4 to the first term on the right side of (9.2.6). □

9.3 Slow Waves

We now prove Proposition 9.2.4.

The idea of the proof is to show that for any $(\vec{a}, \vec{b}) \in \text{BAD} \cap \text{NECC} \cap \text{BB}$, $(\vec{a}, \vec{b})$ satisfies (7.1.7). This, combined with $\text{BAD} \cap \text{NECC} \cap \text{BB}$ implies that:

$$\|\mathcal{P}_{\text{BAD} \cap \text{NECC} \cap \text{BB}} \Psi_0(x)\|_{H_b^s} \leq \delta_{\text{inf}}$$

Thus we need only prove that $(\vec{a}, \vec{b}) \in \text{BAD} \cap \text{NECC} \cap \text{BB}$ satisfies (7.1.7).

We prove first a technical lemma, showing that a given framelet is either incoming or outgoing (not both) if it has velocity sufficiently fast.

Lemma 9.3.1 Assume that w, T_{st} satisfy (7.5.16b) and (7.5.16c). Then for

$$(\vec{a} x_0, \vec{b} k_0) \in [-(L_{\text{in}} + w/3), (L_{\text{in}} + w/3)]^N \times [-k_{\text{sup}}, k_{\text{sup}}]^N,$$

we find that $(\vec{a}, \vec{b}) \notin \text{BAD}(\epsilon, s, T_{\text{st}})$.

Proof By Lemma 6.2.3, it suffices to show that $\mathrm{BB}_{(\vec{a},\vec{b},\sigma)}(\varepsilon,t) \subset [-(L_{\mathrm{in}} + w), (L_{\mathrm{in}} + w)]^N$.

Note that:

$$\vec{a}x_0 + \vec{b}k_0 t \in [-(L_{\mathrm{in}} + w/3 + k_{\sup}t), (L_{\mathrm{in}} + w/3 + k_{\sup}t)]^N$$

$$\subseteq [-(L_{\mathrm{in}} + w/3 + k_{\sup}T_{\mathrm{st}}), (L_{\mathrm{in}} + w/3 + k_{\sup}T_{\mathrm{st}})]^N$$

Consider $\vec{x} \in \mathrm{BB}_{(\vec{a},\vec{b},\sigma)}(\varepsilon,t)$. By Definition 6.2.1 (the definition of $\mathrm{BB}_{(\vec{a},\vec{b},\sigma)}(\varepsilon,t)$), we find that:

$$|\vec{x} - \vec{a}x_0 + \vec{b}k_0 t|_2 \leq \mathfrak{w}_i^s(\vec{b},\epsilon) + \mathfrak{w}_v^s(\vec{b},\epsilon)t$$

Thus, since $\vec{a}x_0 + \vec{b}k_0 t \in [-(L_{\mathrm{in}} + w/3 + k_{\sup}T_{\mathrm{st}}), (L_{\mathrm{in}} + w/3 + k_{\sup}T_{\mathrm{st}})]^N$, and $\mathrm{BB}_{(\vec{a},\vec{b},\sigma)}(\varepsilon,t)$ is contained in a ball of radius $\mathfrak{w}_i^s(\vec{b},\epsilon) + \mathfrak{w}_v^s(\vec{b},\epsilon)T_{\mathrm{st}}$ about $\vec{a}x_0 + \vec{b}k_0 t$, we find that:

$$\mathrm{BB}_{(\vec{a},\vec{b},\sigma)}(\varepsilon,t) \subseteq [-(L_{\mathrm{in}} + w/3 + k_{\sup}T_{\mathrm{st}} + \mathfrak{w}_i^s(\vec{b},\epsilon) + \mathfrak{w}_v^s(\vec{b},\epsilon)T_{\mathrm{st}}),$$

$$(L_{\mathrm{in}} + w/3 + k_{\sup}T_{\mathrm{st}} + \mathfrak{w}_i^s(\vec{b},\epsilon) + \mathfrak{w}_v^s(\vec{b},\epsilon)T_{\mathrm{st}})]^N$$

Then applying (7.5.16b) and (7.5.16c), we find that:

$$[-(L_{\mathrm{in}} + w/3 + k_{\sup}T_{\mathrm{st}} + \mathfrak{w}_i^s(\vec{b},\epsilon) + \mathfrak{w}_v^s(\vec{b},\epsilon)T_{\mathrm{st}}),$$

$$(L_{\mathrm{in}} + w/3 + k_{\sup}T_{\mathrm{st}} + \mathfrak{w}_i^s(\vec{b},\epsilon) + \mathfrak{w}_v^s(\vec{b},\epsilon)T_{\mathrm{st}})]^N \subseteq [-(L_{\mathrm{in}} + w), (L_{\mathrm{in}} + w)]^N$$

Lemma 6.2.3 implies the result we seek.   □

Lemma 9.3.2 *Assume w and T_{st} satisfy (7.5.16b) and (7.5.16c).*
Fix $(\vec{a}, \vec{b}) \in \mathbb{Z}^N \times \mathbb{Z}^N$. Suppose that $(\vec{a}, \vec{b})$ satisfies:

$$\exists j \in 1\dots N, |\vec{a}_j x_0| \geq L_{\mathrm{inf}} \text{ and } \vec{b}_j k_0(\vec{a}_j/|\vec{a}_j|) > k_{\mathrm{inf}} \tag{9.3.1}$$

Suppose also that L_{in}, w and L_{inf} satisfy (7.5.16h), that is $L_{\mathrm{inf}} \geq L_{\mathrm{in}} + w/3$.
Then $(\vec{a}, \vec{b}) \notin \mathrm{NECC}(\epsilon, s, \infty)$.

Proof For the duration of this proof, let j denote the (possibly nonunique) index j for which (9.3.1) holds.

Note that by (7.5.16b) and (7.5.16h), we find that $|\vec{a}_j x_0| \geq L_{\mathrm{in}} + w/3$. For simplicity, suppose that $\vec{a}_j > 0$, and therefore that $\vec{b}_j > 0$.

Then note that:

$$\vec{a}_j x_0 + \vec{b}k_0 t \geq (L_{\mathrm{in}} + w/3) + \mathfrak{w}_v^s(\vec{b},\epsilon)t$$

The constant term was obtained by using (7.5.16h) while the t term was obtained using (7.5.16a).

Thus, we find that:

$$d(\vec{a}x_0 + \vec{b}k_0 t, [-L_{in}, L_{in}]^N) \geq w/3 + \mathfrak{w}_v^s(\vec{b}, \epsilon)t \geq \mathfrak{w}_i^s(vb, \epsilon) + \mathfrak{w}_v^s(\vec{b}, \epsilon)t$$

The last inequality follows by applying (7.5.16b). Applying Lemma 6.2.4 implies that $(\vec{a}, \vec{b}) \notin \text{NECC}$. $\square$

Proof of proposition 9.2.4 We now wish to show that:

$$\left\|(e^{i(1/2)\Delta t} - e^{i(1/2)\Delta_b t})\mathcal{P}_{\text{BAD} \cap \text{NECC} \cap \text{BB}}\Psi_0(x)\right\|_{H_b^s} \leq 2\delta_{\inf} \qquad (9.3.2)$$

We do this by showing that $\text{BAD} \cap \text{NECC} \cap \text{BB}$ is a set which satisfies (7.1.7).

Fix $(\vec{a}, \vec{b}) \in \text{BAD} \cap \text{NECC} \cap \text{BB}$. Note that since $(\vec{a}, \vec{b}) \in \text{BB}$, we find that $|\vec{b}k_0|_\infty \leq k_{\sup}$.

Applying the converse of Lemma 9.3.1, we find that $|\vec{a}x_0|_\infty \geq L_{in} + w/3$.

Now suppose $(\vec{a}, \vec{b})$ satisfies (9.3.1). Then:

$$(\vec{a}, \vec{b}) \notin \text{NECC}(\epsilon, s, \infty) \supseteq \text{BAD} \cap \text{NECC} \cap \text{BB}$$

Thus, if $(\vec{a}, \vec{b}) \in \text{BAD} \cap \text{NECC} \cap \text{BB}$, we find that:

$$!(\exists j \in 1 \ldots N, |\vec{a}_j x_0| \geq L_{inf} \text{ and } \vec{b}_j k_0(\vec{a}_j/|\vec{a}_j|) > k_{inf})$$

This implies that $\text{BAD} \cap \text{NECC} \cap \text{BB}$ is a set satisfying (7.1.7). Hence:

$$\left\|(e^{i(1/2)\Delta t} - e^{i(1/2)\Delta_b t})\mathcal{P}_{\text{BAD} \cap \text{NECC} \cap \text{BB}}\Psi_0(x)\right\|_{H_b^s}$$

$$\leq 2\left\|\mathcal{P}_{\text{BAD} \cap \text{NECC} \cap \text{BB}}\Psi_0(x)\right\|_{H_b^s} \leq 2\delta_{\inf}$$

Thus, we have proved Proposition 9.2.4. $\square$

9.4   Estimates of $Q(t)$

We now attempt to determine bounds on $Q(t)$ and $Q_b(t)$ based on apriori and aposteriori knowledge of $\Psi(x, t)$ and $g(t, \vec{x}, \cdot)$. This is where we use Assumption 4.

The main tool is phase space localization based on the WFT and Assumption 4. In particular, we wish to treat $g(t, \vec{x}, \Psi(t))\Psi(t)$ as a source term and then figure out how much of it's mass can leave $[-L_{in}, L_{in}]^N$. We will decompose $\mathbb{Z}^N \times \mathbb{Z}^N = NL \cup NL^C$ (with the set NL defined as in Assumption 4), and write:

$$g(t, \vec{x}, \Psi)\Psi = \sum_{(\vec{a},\vec{b}) \in NL} g_{(\vec{a},\vec{b})}(t)\phi_{(\vec{a},\vec{b})}(\vec{x}) + \sum_{(\vec{a},\vec{b}) \in NL^C} g_{(\vec{a},\vec{b})}(t)\phi_{(\vec{a},\vec{b})}(\vec{x})$$

The last term is small by Assumption 4. We now come up with sufficient conditions on L_{in} and T_{st} (depending on $k_{sup,NL}$ and L_{NL}) so that framelets in NL are not bad.

Proposition 9.4.1 *Let* T_{st}, L_{in} *satisfy* (7.5.16e), (7.5.16f) *and* (7.5.16g). *Then* $NL \cap BAD(\epsilon, s, T_{st}) = \emptyset$.

Proof Fix $(\vec{a}, \vec{b}) \in NL$.

Note that $BB_{(\vec{a},\vec{b},\sigma)}(\epsilon, t)$ is a ball of radius $\mathfrak{w}_i^s(\vec{b}, \epsilon) + \mathfrak{w}_v^s(\vec{b}, \epsilon)t$ around the point $\vec{a}x_0 + \vec{b}k_0 t$. Thus, if $\vec{x} \in BB_{(\vec{a},\vec{b},\sigma)}(\epsilon, t)$, then:

$$|\vec{x}|_\infty \leq |\vec{a}x_0|_\infty + |\vec{b}k_0|_\infty t + \mathfrak{w}_i^s(\vec{b}, \epsilon) + \mathfrak{w}_v^s(\vec{b}, \epsilon)t$$

$$\leq L_{in} + k_{sup,NL}T_{st} + w/2 + \mathfrak{w}_v^s(\vec{b}, \epsilon)T_{st}$$

$$\leq L_{in} + w/2 + (L_{in} + w/2 - L_{NL})$$

This calculation follows by applying (7.5.16e) to $(k_{sup,NL} + \mathfrak{w}_v^s(\vec{b}, \epsilon))T_{st}$ and (7.5.16g) to $\mathfrak{w}_i^s(\vec{b}, \epsilon)$.

Note that (7.5.16f) is needed only to insure that (7.5.16e) is possible to satisfy, i.e. that $L_{in} - L_{NL} > 0$.

This implies that $\vec{x} \in [-(L_{in} + w), (L_{in} + w)]^N$, hence

$$BB_{(\vec{a},\vec{b},\sigma)}(\epsilon, t) \subset [-(L_{in} + w), (L_{in} + w)]^N$$

Applying Lemma 6.2.3 implies that $(\vec{a}, \vec{b}) \notin BAD(\epsilon, s, T_{st})$. The only assumption on $(\vec{a}, \vec{b})$ was $(\vec{a}, \vec{b}) \in NL$, hence $NL \cap BAD(\epsilon, s, T_{st}) = \emptyset$. $\square$

We can now compute a bound on $Q(t)$ for $Q(t)$ satisfying Assumption 4.

Proposition 9.4.2 *Let* $g(t, \vec{x}, \Psi)\Psi$ *satisfy Assumption 4. Suppose that* L_{in} *and* T_{st} *satisfy* (7.5.16e), (7.5.16f) *and* (7.5.16g). *Then* $Q(t)$ *satisfies:*

$$Q(t) \leq (\widehat{Q}(t)\mathcal{G} + t\delta_{NL}) \sup_{t' \in [0,t]} \|\Psi(x, t')\|_{H^s} \tag{9.4.1}$$

Proof We note that:

$$\left\| \int_0^t (e^{i(1/2)\Delta(t-t')} - e^{i(1/2)\Delta_b(t-t')}) g(t,\vec{x},\Psi(\vec{x},t'))\Psi(\vec{x},t')dt' \right\|_{H_b^s}$$

$$\leq \int_0^t \left\| (e^{i(1/2)\Delta(t-t')} - e^{i(1/2)\Delta_b(t-t')}) \sum_{(\vec{a},\vec{b})\in NL} g_{(\vec{a},\vec{b})}(t)t')\phi_{(\vec{a},\vec{b})}(\vec{x}) \right\|_{H_b^s} dt'$$

$$+ \int_0^t \left\| (e^{i(1/2)\Delta(t-t')} - e^{i(1/2)\Delta_b(t-t')}) \sum_{(\vec{a},\vec{b})\in NL^C} g_{(\vec{a},\vec{b})}(t)t')\phi_{(\vec{a},\vec{b})}(\vec{x}) \right\|_{H_b^s} dt'$$

By Assumption 4, for any fixed t, the last term satisfies:

$$\left\| (e^{i(1/2)\Delta(t-t')} - e^{i(1/2)\Delta_b(t-t')}) \sum_{(\vec{a},\vec{b})\in NL^C} g_{(\vec{a},\vec{b})}(t)t')\phi_{(\vec{a},\vec{b})}(\vec{x}) \right\|_{H_b^s}$$

$$\leq 2\delta_{NL}t \sup_{t'\in[0,t]} \left\| \Psi(x,t') \right\|_{H^s} \qquad (9.4.2)$$

The first term satisfies (at each fixed $t \leq T_{st}$):

$$\left\| (e^{i(1/2)\Delta(t-t')} - e^{i(1/2)\Delta_b(t-t')}) \sum_{(\vec{a},\vec{b})\in NL} g_{(\vec{a},\vec{b})}(t)t')\phi_{(\vec{a},\vec{b})}(\vec{x}) \right\|_{H_b^s}$$

$$\leq \left\| g(t,\vec{x},\Psi(\vec{x},t))\Psi(\vec{x},t) \right\|_{L^2} \mathcal{A}_F^{-1} \sqrt{\sum_{(\vec{a},\vec{b})\in NL} \left| \mathfrak{E}_{(\vec{a},\vec{b})}(t) \right|^2}$$

$$\leq \mathcal{G} \left\| \Psi(\vec{x},t) \right\|_{H^s} \mathcal{A}_F^{-1} \sqrt{\sum_{(\vec{a},\vec{b})\in NL} \left| \mathfrak{E}_{(\vec{a},\vec{b})}(t) \right|^2}$$

We then integrate this result over time:

$$
\int_0^t \mathcal{G} \, \|\Psi(\vec{x}, t)\|_{H^s} \, \mathcal{A}_F^{-1} \sqrt{\sum_{(\vec{a}, \vec{b}) \in \mathrm{NL}} \left| \mathfrak{E}_{(\vec{a}, \vec{b})}(t) \right|^2}
$$

$$
\leq \mathcal{G}t \sup_{t' \in [0,t]} \|\Psi(\vec{x}, t)\|_{H^s} \, \mathcal{A}_F^{-1} \sqrt{\sum_{(\vec{a}, \vec{b}) \in \mathrm{NL}} \left| \mathfrak{E}_{(\vec{a}, \vec{b})}(t) \right|^2} = \mathcal{G} \, \|\Psi(\vec{x}, t)\|_{H^s} \, \widehat{\mathcal{Q}}(t)
$$

$$
(9.4.3)
$$

Adding (9.4.2) and (9.4.3) yields the result we seek.   □

Chapter 10
Proof of Theorems 7.5.4 and 7.5.5

In this part, we prove Theorems 7.5.4 and 7.5.5. We first prove a technical result, that the waves outside $[-(L_{in} + w), (L_{in} + w)]^N$ and also the waves of high frequency are small. We will use this result in the proof of both Theorems 7.5.4 and 7.5.5. Now we will show that $f(x)$ is localized. We assume throughout this section that $f(x) = \mathfrak{U}(T_{st})\mathcal{P}_{NECC \cap BB}h(x)$ for some $h(x) \in H^s$.

Proposition 10.0.1 *The following inequality holds.*

$$\|f(x)\|_{H^s(\mathbb{R}^N \setminus [-L_{trunc}, L_{trunc}]^N)} \leq \widehat{\mathcal{E}}(T_{st}) \|h(x)\|_{L^2} + \widehat{\mathcal{Q}}(T_{st}) \tag{10.0.1}$$

where $E(t)$ and $Q(t)$ are given by (7.5.18a) and (7.5.18b) with $h(x)$ replacing $\Psi(x)$.

Proof Recall that the per-framelet error functions, used to construct $\widehat{\mathcal{E}}(t) \|h(x)\|_{L^2}$ and $\widehat{\mathcal{Q}}(t)$ are nothing more than the mass (in H^s) outside $[-(L_{in}+w), (L_{in}+w)]^N$. Thus, the proofs of Propositions 9.2.5 and 9.4.2 apply without change, and we can merely add $\widehat{\mathcal{E}}(T_{st}) \|h(x)\|_{L^2}$ and $\widehat{\mathcal{Q}}(T_{st})$ to get our bound. □

Proposition 10.0.2 *The framelet coefficients of $f(x)$ satisfy:*

$$\left\|\mathcal{P}_{BB^C} f(x)\right\|_{H^s} \leq \mathfrak{H}^s_+(\tilde{g}(\vec{x}))\mathfrak{H}^{-s}_+(e^{-x^2/\sigma^2})[\widehat{\mathcal{E}}(T_{st}) \|h(x)\|_{L^2} + \widehat{\mathcal{Q}}(T_{st}) + \epsilon] + \delta_{sup} \tag{10.0.2}$$

Proof Note that BB^C Cconsists of framelets moving faster than k_{sup}, or outside the region $[-L_{trunc}, L_{trunc}]^N$. We apply Corollary 3.3.9.

Assumption 2 can be invoked to bound $\left\|\mathcal{P}_{HF(k_{sup})} f(x)\right\|_{H^s}$ by δ_{sup}. To bound the spatial component, we apply Proposition 10.0.1.

$$\|\mathcal{P}_{BB} f(x)\|_{H^s} \leq \mathfrak{H}^s_+(\tilde{g}(\vec{x}))\mathfrak{H}^{-s}_+(e^{-x^2/\sigma^2})[E(T_{st}) + Q(T_{st}) + \epsilon]$$

$$+ \left\|\mathcal{P}_{HF(k_{sup})} f(x)\right\|_{H^s} \tag{10.0.3}$$

□

© The Author(s), under exclusive license to Springer Nature Singapore Pte Ltd. 2023
A. Soffer et al., *Time Dependent Phase Space Filters*, SpringerBriefs on PDEs and Data Science, https://doi.org/10.1007/978-981-19-6818-1_10

10.1   Outgoing Waves

In this section we prove Theorem 7.5.4, concerning the outgoing wave term:

$$\left\| \mathfrak{U}(t|f) \mathcal{P}_{\text{NECC}^C} f(x) \right\|_{H_b^s}$$

Our goal is to show that because the waves are outgoing, this term remains small for a long time. The function $f(x)$ will be assumed to satisfy Assumption 2, and also satisfy the assumption that $f(x) = \mathfrak{U}(t) \mathcal{P}_{\text{NECC} \cap \text{BB}} h(x)$.

This is where we use Assumption 5. Assumption 5 states that:

$$\left\| \mathfrak{U}(t|f) \mathcal{P}_{\text{NECC}^C} f(x) \right\|_{H_b^s} \leq \delta_{\text{F}} \left\| \mathcal{P}_{\text{NECC}^C} f(x) \right\|_{H_b^s}$$

We first add and subtract $e^{i(1/2)\Delta t} \mathcal{P}_{\text{NECC}} f$ under the norm, and apply the triangle inequality:

$$\left\| \mathfrak{U}(t|f) \mathcal{P}_{\text{NECC}^C} f(x) \right\|_{H_b^s}$$

$$\leq \left\| \mathfrak{U}(t) \mathcal{P}_{\text{NECC}^C} f(x) - e^{i(1/2)\Delta t} \mathcal{P}_{\text{NECC}^C} f(x) \right\|_{H_b^s}$$

$$+ \left\| e^{i(1/2)\Delta t} \mathcal{P}_{\text{NECC}^C} f(x) \right\|_{H_b^s} \tag{10.1.1}$$

The first term is bounded by $\delta_{\text{F}} \left\| \mathcal{P}_{\text{NECC}^C} f(x) \right\|_{H^s}$, by Assumption 5, combined with (7.5.7b). The constraint (7.5.7b) implies that a sphere of radius $\mathfrak{W}^+(\vec{b}, \delta_{\text{F}}, t)$ about $\vec{a} x_0 + \vec{b} k_0 t$ is contained entirely in $\text{BB}_{(\vec{a}, \vec{b}, \sigma)}(\epsilon, t)$, and therefore does not intersect $[-L_{\text{in}}, L_{\text{in}}]^N \supset [-L_F, L_F]^N$.

Thus we need only compute a bound on $\left\| e^{i(1/2)\Delta t} \mathcal{P}_{\text{NECC}^C} f(x) \right\|_{H_b^s}$. We break up $\mathcal{P}_{\text{NECC}^C} f(x)$ further:

$$\mathcal{P}_{\text{NECC}^C} f(x) = \mathcal{P}_{\text{NECC}^C \cap \text{HF}(k_{\text{sup}}) \cup B} f(x) + \mathcal{P}_{\text{NECC}^C \cap (\text{HF}(k_{\text{sup}}) \cup B)^C} f(x)$$

Proposition 10.0.2 provides a bound on the first term. To bound the second, we need merely count the framelets in $(\text{HF}(k_{\text{sup}}) \cup B)^C$ and apply Lemma 9.2.1.

We observe that B^C consists only of framelets with $|\vec{a}|_\infty x_0 \leq L_{\text{trunc}} + \mathfrak{X}^s(\epsilon, k_{\text{sup}})$, while $\text{HF}(k_{\text{sup}})^C$ consists only of framelets with $\left| \vec{b} \right|_\infty x_0 \leq k_{\text{sup}}$. It is easy to see that there are only $(2k_{\text{sup}}/k_0)^N (2[L_{\text{trunc}} + \mathfrak{X}^s(\epsilon, k_{\text{sup}})]/x_0)^N$ such framelets. Thus, we obtain the result of Theorem 7.5.4:

$$\left\| \mathfrak{U}(t|f) \mathcal{P}_{\text{NECC}^C} f(x) \right\|_{H_b^s} \leq \delta_{\text{F}} \left\| \mathcal{P}_{\text{NECC}^C} f(x) \right\|_{H_b^s}$$

$$+ \mathfrak{H}_+^s (\tilde{g}(\vec{x})) \mathfrak{H}_+^{-s} (e^{-x^2/\sigma^2})[E(T_{\text{st}}) + Q(T_{\text{st}}) + \epsilon] + \delta_{\text{sup}}$$

$$+ \mathcal{A}_F^{-1} \left(\sum_{(\vec{a},\vec{b}) \in \text{NECC}^C \cap (\text{HF}(k_{\sup}) \cup B)^C} \left| \mathcal{R}_{(\vec{a},\vec{b})}(s)t \right|^2 \right)^{1/2} \|f(x)\|_{L^2}$$

$$= \delta_{\text{F}} \left\| \mathcal{P}_{\text{NECC}^C} f(x) \right\|_{H_b^s} + \widehat{\mathcal{R}}(t) \|f(x)\|_{L^2}$$

$$+ \mathfrak{H}_+^s(\tilde{g}(\vec{x})) \mathfrak{H}_+^{-s}(e^{-x^2/\sigma^2})[\widehat{\mathcal{E}}(T_{\text{st}}) \|h(x)\|_{L^2} + \widehat{\mathcal{Q}}(T_{\text{st}}) + \epsilon] + \delta_{\sup} \qquad (10.1.2)$$

Here, we used definition of $\widehat{\mathcal{R}}(t)$ to simplify the inequality.

10.2 Residual Waves

In this section, we wish to show that

$$\left\| \mathfrak{U}(t)\mathcal{P}_{\text{NECC}} f(x) - \mathfrak{U}(t)\mathcal{P}_{\text{NECC} \cap \text{BB}} f(x) \right\|_{H_b^s} = \left\| \mathfrak{U}(t)\mathcal{P}_{\text{NECC} \setminus \text{BB}} f(x) \right\|_{H_b^s}$$

is small, provided $f(x) = \Phi(x, nT_{\text{st}})$ for some n.

The residual waves consist of waves which are located outside the box, but are moving in a direction that will take them into the box at some future point. They can be thought of as outgoing waves that have turned around outside the box, and are returning.

Remark 10.2.1 This proof does not use the fact that the waves are off the propagation set. It merely uses the fact that BB^C consists of framelets which are localized outside the box, and it takes a moderate amount of time to reach them.

Proof of Theorem 7.5.5 By proposition 10.0.2, and the observation that

$$\text{NECC} \setminus \text{BB} \subset \text{BB}^C$$

we observe that:

$$\left\| \mathcal{P}_{\text{NECC} \setminus \text{BB}} f(x) \right\|_{H^s}$$

$$\leq \mathfrak{H}_+^s(\tilde{g}(\vec{x})) \mathfrak{H}_+^{-s}(e^{-x^2/\sigma^2})[E(T_{\text{st}}) + Q(T_{\text{st}}) + \epsilon]$$

$$+ \left\| \mathcal{P}_{\text{HF}(k_{\sup})} f(x) \right\|_{H^s}$$

We then observe that:

$$\left\| \mathfrak{U}(t|\mathcal{P}_{\text{NECC}} f)\mathcal{P}_{\text{NECC}} f(x) - \mathfrak{U}(t|\mathcal{P}_{\text{NECC} \cap \text{BB}})\mathcal{P}_{\text{NECC} \cap \text{BB}} f(x) \right\|_{H_b^s}$$

$$\leq \mathcal{L}(t) \left\| \mathcal{P}_{\text{NECC}} f(x) - \mathcal{P}_{\text{NECC} \cap \text{BB}} f(x) \right\|_{H_b^s}$$

$$\leq \mathcal{L}(t)\left(\mathfrak{H}_+^s(\tilde{g}(\vec{x}))\mathfrak{H}_+^{-s}(e^{-x^2/\sigma^2})[E(T_{\mathrm{st}}) + Q(T_{\mathrm{st}}) + \epsilon]\right.$$

$$\left. + \left\|\mathcal{P}_{\mathrm{HF}(k_{\mathrm{sup}})} f(x)\right\|_{H^s}\right)$$

This is the result we seek, after substituting the definitions of $E(T_{\mathrm{st}})$ and $Q(T_{\mathrm{st}})$ in. □

Chapter 11
Numerical Experiments

In this part, we discuss the results of our numerical tests. The code is based on Python 3.11. It is available for download from the last author's webpage: http://minhbinhtran.org/python/ The code is being extended so that it works several types of wave equations [52]. For example, this is a simulation for the Euler equation and this is a simulation for the Maxwell equation.

11.1 Case 1: $T + R = E$

The standard method for testing absorbing boundaries is simply to throw coherent states (which are well localized in frequency) at the boundary and compute transmission (T) and reflection (R) coefficients. The sum $T + R$ is then taken to be the error associated to that particular absorbing boundary condition.

This is a useful test, although it is by no means completely characterizes the errors (see Sect. 11.2). We performed the following numerical experiment. The free Schrödinger equation (i.e. $g(t, \vec{x}, \Psi(x, t))\Psi(x, t) = 0$) was solved on the box $[-25.6, 25.6]$ with TDPSF boundaries on the region $[-25.6, -12]$ and $[12, 25.6]$. The initial condition was taken to be $\Psi(x, 0) = e^{-x^2/4}e^{ivx}$, with v ranging from 1 to 25. After the initial condition was given sufficient time to wrap around the computational domain, the mass inside the region $[-10, 10]$ was measured, and the result is graphed in Fig. 11.1. This was done with the WFT frame using $\sigma = 1, 2, 4$. The results are comparable to the complex absorbing potential $V(x) = -25ie^{-(x-25.6)^2/16}$, which is also shown in Fig. 11.1. The width of the complex potential was chosen so that it's spatial extent is comparable to the width of the TDPSF used.

This particular example demonstrates no major advantage of the TDPSF over the absorbing potential. The TDPSF works better for some velocities, but not all. The advantage of the TDPSF is not that it succesfully dissipates outgoing waves. The

© The Author(s), under exclusive license to Springer Nature Singapore Pte Ltd. 2023
A. Soffer et al., *Time Dependent Phase Space Filters*, SpringerBriefs on PDEs and Data Science, https://doi.org/10.1007/978-981-19-6818-1_11

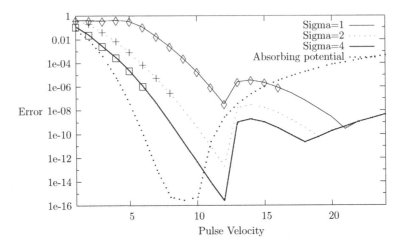

Fig. 11.1 A simulation of $T + R$ vs the velocity of an outgoing pulse. Our numerical data are comparable to an absorbing potential

advantage of the TDPSF is that it does not dissipate incoming waves. An example of this will be demonstrated in the next section.

11.2   Case 2: $T + R \neq E$

We describe in this section a scenario in which computing a bound on $T+R$ provides no useful estimate.

Consider the following linear Schrödinger equation (with $(\vec{x}, t) \in \mathbb{R}^{2+1}$):

$$i \partial_t \Psi(x, t) = \left[-(1/2)\Delta - \frac{15}{0.05|\vec{x}|^2 + 1} \right] \Psi(x, t) \tag{11.2.1}$$

$$\Psi(x, 0) = e^{i7x_2} e^{-|\vec{x}|^2/20} + e^{i4x_1} e^{-|\vec{x}|^2/20}$$

Observe that the initial condition consists of two coherent states of equal mass, one with velocity 4 and one with velocity 7. The notable fact about this particular potential is that the fast gaussian has enough kinetic energy to (mostly) escape from the binding potential. The slow gaussian does not. The slow gaussian moves toward the boundary, turns around and returns.

The problem with the absorbing potential approach is that the absorbing potential does not distinguish between incoming and outgoing waves. It dissipates everything on the boundary including the waves that should have returned. This will occur even if one can construct a complex potential for which $T + R = 0$!

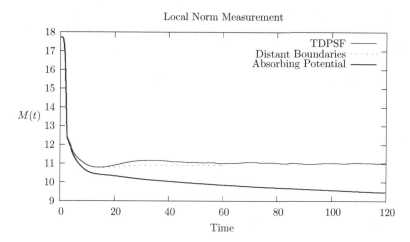

Fig. 11.2 A picture of $M(t) = \|\Psi(x, t)\|_{L^2([-10, 10]^2)}$. As the outgoing pulse returns at $t = 29$, the distant boundary simulation is invalid at time

We ran three simulations of (11.2.1). The first was performed using the TDPSF with $\sigma = 2.0$. The region of computation was $[-25.6, 25.6]^2$ The second was performed (on the same region) with an absorbing potential

$$V_1(\vec{x}) = -20i\,e^{-(\tilde{x}_1 \pm 25.6)^2/36} - 20i\,e^{-(\tilde{x}_2 \pm 25.6)^2/36}.$$

The third was solved with periodic boundary conditions on the region $[-102.4, 102.4]^2$. This boundary is sufficiently distant so that the outgoing waves cannot return to the origin for a time $204.8/7.0 \approx 29$. Thus, we will take the distant boundaries simulation as our benchmark, at least for $t \leq 29$.

After $t = 29$, we have some qualitative knowledge of the behavior. We expect that the solution consists of continuum and bound states. Over a short time, the continuum will disperse, leaving only the bound states. The bound states will remain forever.

In all three cases, the quantity $M(t) = \|\Psi(x, t)\|_{L^2([-10, 10]^2)}$ was computed. The simulation using the TDPSF agreed with the simulation on the larger region to within 1.25% for $t < 29$.[1] The simulation using complex potentials had an error of 4% for $t < 29$, and the error appears to increase after that.

In fact, examining the graphs of $M(t)$ (see Fig. 11.2) part of the bound states appear to be dissipating. In fact, we believe that this dissipation will continue and the error will only get worse with time.

---

[1] In fact, the 1.25% is much better than one might otherwise expect. A simple calculation shows that the potential is equal to -0.44 on the boundary. Therefore, Assumption 4 is not satisfied, since the "nonlinearity" is not contained inside the box. Additional simulations using the domain $[-51.2, 51.2]^2$ yielded almost complete agreement with the simulation using distant boundaries, and had the correct qualitative behavior after that.

The reason the TDPSF performs so much better than the complex potential is that it distinguishes outgoing waves from incoming waves on the boundary. The TDPSF only removes waves which sit on the boundary and are also outgoing with sufficiently high velocity. The trapped waves, although they sit on the boundary, do not have high outgoing velocity, and thus are not removed.

11.3 On Assumption 4

Consider the nonlinearity, $g(t, \vec{x}, \Psi(\vec{x}, t)) = -|\Psi(\vec{x}, t)|^2$. It is desirable to construct a numerical algorithm which filters outgoing solitons as well as free waves. Although the TDPSF was not designed to filter solitons, it turns out to work well for solitons which are not moving slowly.

The reason for this is that an outgoing soliton with sufficiently high velocity is localized in phase space on outgoing waves. Consider a simple soliton, $\phi(x, v, t) = 2^{-1/2} e^{i(vx + (1-v^2)t)} / \cosh((x - vt))$. A simple calculation shows that for $|k - v| \gg 1$, $\hat{\phi}(k, v, t) \sim e^{-|k-v|}$. Thus, for $k \gg k_{inf}$, this shows that the framelet coefficients of $\phi(x, v, t)$ which are moving too slowly to resolve have exponentially small mass. This shows that under the free flow this soliton is strictly outgoing.

The soliton is also leaving the box under the full flow $\mathfrak{U}(t)$. Although $e^{i(1/2)\Delta t}$ and $\mathfrak{U}(t)$ move the soliton very differently (one dispersively, one coherently), they both move it out of the box and in nearly the same direction. For this reason, the TDPSF is expected to filter soliton solutions correctly.

We ran numerical tests to demonstrate this as follows. We solved the (1.1.1) with $g(t, \vec{x}, \Psi(\vec{x}, t)) = -|\Psi(\vec{x}, t)|^2$ on the region $[-25.6, 25.6]$ with $\delta x = 0.05$, $\delta t = 0.002/v$ and $T_{st} = 0.008/v$ (the timestep's are scaled with the velocity to speed up the simulations). In this simulation, $L_{in} = 12.0$ and $w = 13.6$. The initial condition was taken to be $\Psi(x, 0) = 2^{-1/2} e^{ivx} / \cosh(x)$ for $v = 1..15$.

The TDPSF was used with $T_{st} = 0.08/v$, $x_0 = 0.20$, $k_0 = 2\pi/3.2$, and $\sigma = 1, 2, 3$. We measured the following quantity:

$$E(v) = \sup_{t < 200/v} \frac{\|\Psi(x, t) - \Psi_{ex}(x, t)\|_{L^2([-10, 10])}}{\|\Psi_{ex}(x, 0)\|_{L^2(\mathbb{R})}} \tag{11.3.1}$$

The function $\Psi_{ex}(x, t)$ is the exact solution. The result of this experiment is plotted in Fig. 11.3. The time $200/v$ was chosen since it is more than enough time for errors to return to the region $[-10, 10]$. We believe the error floor near 10^{-10} visible in Fig. 11.3 is due to underlying machine errors, as well as certain approximations we made when computing the WFT.

We discuss a possible better way to filter solitons in Sect. 11.3.2.

Remark 11.3.1 The paper [67] proposes an alternative method of absorbing boundaries (namely the paradifferential strategy), based on a novel method of approximating the Dirichlet-to-Neumann operator. A similar numerical test was performed for

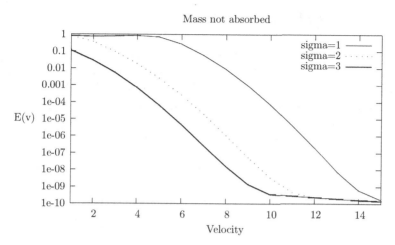

Fig. 11.3 A graph of the error (see (11.3.1)) as a function of velocity. We observe the exponential improvement in accuracy with velocity until the error reaches 10^{-10}

those boundary conditions. For a soliton at velocity 15, Szeftel obtained $E(15) = 0.08$ at best. For comparison, we obtain $E(15) = 1.69 \times 10^{-10}$ for $\sigma = 1$ and $E(v) = 1.40 \times 10^{-10}$ for $\sigma = 3$.

Our region of interest was $[-12, 12]$ with TDPSF region on $[-25.6, -12]$ and $[12, 25.6]$, as opposed to Szeftel who used $[-5, 5]$. We used FFT/Split Step propagation for the interior problem, as opposed to the finite elements of [67]. However, the comparison in this case is focused on the behavior of the absorbing boundary conditions for a moving soliton. Since the soliton of this equation is essentially constant in form, the size of the domain along which it moved, until reaching the filter on boundary is not relevant. So, the comparison is valid for this initial data. And what we see is that the filter we use is more accurate by 9 orders of magnitude. In addition, it is quite surprising that the TDPSF differs from the Dirichlet-to-Neumann boundary by such a large magnitude, because [67] actually takes the nonlinearity into account while the TDPSF assumes the nonlinearity is zero on the boundary.

Our tools were not designed to filter an outgoing soliton or other coherent structures. The case of soliton is different and requires a modification of the filter. A powerful way for that is rewriting the original equation as a system of equations called modulation equations. Then the filter works as usual on the PDE part, and the ODE, describing the motion of the soliton is done separately. A soliton cannot be filtered by an outgoing wave decomposition. Soliton is a clump of both incoming and outgoing waves. However, if the speed of the soliton is sufficiently large compared to its velocity spread, then most components are effectively outgoing, and can be filtered.

11.3.1 A Coincidence

This is not a general phenomenon, however, as illustrated by the following example. Consider the KdV equation, solved by a comparable TDPSF algorithm. That is, we decompose the solution into framelets, and filter those which are leaving B under the free flow $e^{\partial_x^3}$. A soliton near the right boundary will be filtered since it is leaving under the free flow. But under the full flow, the soliton will not leave the box, since solitons propagate leftward while free waves propagate rightward.

The fact that our method succesfully filters outwardly moving solitons is a consequence of the fact that fast-moving solitons for the NLS have a small projection onto incoming waves. For some nonlinearities, a soliton or soliton-like object at position $(\vec{a}x_0, \vec{b}k_0)$ in phase space actually propagates along the trajectory $\vec{a}x_0 + t\vec{b}k_0$. If the dynamics of the equation lack this property, there is no reason to believe a TDPSF algorithm will effectively filter solitons.

11.3.2 Motivation in Constructing the TDPSF

Our motivation in constructing the TDPSF was the following. Nonreflecting boundary conditions are possible because we understand the motion of waves away from the support of the nonlinearity. We used that knowledge to determine what to filter, and what not to filter.

We propose that a practical way to filter outgoing solitons is simply to identify them and remove them. That is, at a time T_{st}, we determine whether $\Psi(x, T_{st})$ might have a soliton located near the boundary. If so, use the decomposition $\Psi(x, T_{st}) = S(x) + R(x)$, where $S(x)$ is the soliton and $R(x)$ is the remainder. We then determine whether $S(x)$ is outgoing. If it is, we then set $\Psi(x, T_{st+}) = R(x)$, filtering the soliton. This does depend on an explicit knowledge of what solitons look like. That information is often available, so this assumption is not unreasonable.

References

1. M. Abramawitz and I.A. Stegun. *Handbook of Mathematical Functions*. Dover, 1965.
2. X. Antoine, A. Arnold, C. Besse, M. Ehrhardt, and A. Schädle. A review of transparent and artificial boundary conditions techniques for linear and nonlinear schrödinger equations. 2008.
3. X. Antoine, W. Bao, and C. Besse. Computational methods for the dynamics of the nonlinear schrödinger/gross–pitaevskii equations. *Computer Physics Communications*, 184(12):2621–2633, 2013.
4. X. Antoine, C. Besse, and S. Descombes. Artificial boundary conditions for one-dimensional cubic nonlinear schrödinger equations. *SIAM journal on numerical analysis*, 43(6):2272–2293, 2006.
5. X. Antoine, C. Besse, and P. Klein. Absorbing boundary conditions for the one-dimensional schrödinger equation with an exterior repulsive potential. *Journal of Computational Physics*, 228(2):312–335, 2009.
6. X. Antoine, C. Besse, and P. Klein. Absorbing boundary conditions for general nonlinear schrodinger equations. *SIAM Journal on Scientific Computing*, 33(2):1008–1033, 2011.
7. X. Antoine, C. Besse, and P. Klein. Numerical solution of time-dependent nonlinear schrödinger equations using domain truncation techniques coupled with relaxation scheme. *Laser Physics*, 21(8):1491–1502, 2011.
8. X. Antoine, C. Besse, and P. Klein. Absorbing boundary conditions for the two-dimensional schrödinger equation with an exterior potential part i: construction and a priori estimates. *Mathematical Models and Methods in Applied Sciences*, 22(10):1250026, 2012.
9. X. Antoine, C. Besse, and P. Klein. Absorbing boundary conditions for the two-dimensional schrödinger equation with an exterior potential. *Numerische Mathematik*, 125(2):191–223, 2013.
10. X. Antoine, C. Geuzaine, and Q. Tang. Perfectly matched layer for computing the dynamics of nonlinear schrödinger equations by pseudospectral methods. application to rotating bose-einstein condensates. *Communications in Nonlinear Science and Numerical Simulation*, 90:105406, 2020.
11. X. Antoine, E. Lorin, J. Sater, F. Fillion-Gourdeau, and A. D. Bandrauk. Absorbing boundary conditions for relativistic quantum mechanics equations. *Journal of Computational Physics*, 277:268–304, 2014.
12. X. Antoine, E. Lorin, and Q. Tang. A friendly review of absorbing boundary conditions and perfectly matched layers for classical and relativistic quantum waves equations. *Molecular Physics*, 115(15–16):1861–1879, 2017.
13. A. Arnold. Numerically absorbing boundary conditions for quantum evolution equations. *VLSI design*, 6(1–4):313–319, 1998.

14. A. Arnold, N. B. Abdallah, and C. Negulescu. Wkb-based schemes for the oscillatory 1d schrödinger equation in the semiclassical limit. *SIAM journal on numerical analysis*, 49(4):1436–1460, 2011.

15. C. Bardos, L. Halpern, G. Lebeau, J. Rauch, and E. Zuazua. Stabilisation de l'équation des ondes au moyen d'un feedback portant sur la condition aux limites de dirichlet. *Asymptotic analysis*, 4(4):285–291, 1991.

16. A. Bayliss and E. Turkel. Radiation boundary conditions for wave-like equations. *Communications on Pure and applied Mathematics*, 33(6):707–725, 1980.

17. S. Becker, J. Sewell, and E. Tebbutt. Computability of magnetic Schrödinger and Hartree equations on unbounded domains. *Numerical Methods for Partial Differential Equations*, 39(2), 1299–1332, 2023.

18. J.-P. Berenger. A perfectly matched layer for the absorption of electromagnetic waves. *J. Comput. Phys.*, 114(2):185–200, 1994.

19. C. Besse, B. Mésognon-Gireau, and P. Noble. Artificial boundary conditions for the linearized benjamin–bona–mahony equation. *Numerische Mathematik*, 139(2):281–314, 2018.

20. J. M. Bony. Calcul symbolique et propagation des singularites pour les equations aux derivees partielles non lineaires. *Ann. Sci. Ec. Norm. Sup (4 eme serie)*, 14:209–246.

21. J. Bourgain. *Global solutions of nonlinear Schrödinger equations*, volume 46 of *American Mathematical Society Colloquium Publications*. American Mathematical Society, Providence, RI, 1999.

22. E. Candès and L. Demanet. Curvelets and Fourier integral operators. *C. R. Math. Acad. Sci. Paris*, 336(5):395–398, 2003.

23. E. J. Candès and L. Demanet. The curvelet representation of wave propagators is optimally sparse. *Comm. Pure Appl. Math.*, 58(11):1472–1528, 2005.

24. R. M. Caplan and R. Carretero-González. A modulus-squared dirichlet boundary condition for time-dependent complex partial differential equations and its application to the nonlinear schrodinger equation. *SIAM Journal on Scientific Computing*, 36(1):A1–A19, 2014.

25. T. Cazenave. *Semilinear Schrödinger equations*, volume 10 of *Courant Lecture Notes in Mathematics*. New York University Courant Institute of Mathematical Sciences, New York, 2003.

26. H. L. Cycon, R. G. Froese, W. Kirsch, and B. Simon. *Schrödinger operators with application to quantum mechanics and global geometry*. Texts and Monographs in Physics. Springer-Verlag, Berlin, 2009.

27. I. Daubechies. The wavelet transform, time-frequency localization and signal analysis. *IEEE Trans. Inform. Theory*, 36(5):961–1005, 1990.

28. I. Daubechies. *Ten lectures on wavelets*, volume 61 of *CBMS-NSF Regional Conference Series in Applied Mathematics*. Society for Industrial and Applied Mathematics (SIAM), Philadelphia, PA, 1992.

29. I. Daubechies and A. Grossmann. Frames in the Bargmann space of entire functions. *Comm. Pure Appl. Math.*, 41(2):151–164, 1988.

30. I. Daubechies, A. Grossmann, and Y. Meyer. Painless nonorthogonal expansions. *J. Math. Phys.*, 27(5):1271–1283, 1986.

31. B. Engquist and L. Halpern. Long-time behaviour of absorbing boundary conditions. *Mathematical methods in the applied sciences*, 13(3):189–203, 1990.

32. B. Engquist and A. Majda. Radiation boundary conditions for acoustic and elastic wave calculations. *Comm. Pure Appl. Math.*, 32(3):314–358, 1979.

33. Bj. Engquist and A. Majda. Absorbing boundary conditions for numerical simulation of waves. *Proc. Nat. Acad. Sci. U.S.A.*, 74(5):1765–1766, 1977.

34. S. Ervedoza and E. Zuazua. Perfectly matched layers in 1-d: Energy decay for continuous and semi-discrete waves. *Numerische Mathematik*, 109(4):597–634, 2008.

35. D. Givoli. Non-reflecting boundary conditions. *Journal of computational physics*, 94(1):1–29, 1991.

36. M.s J. Grote and C. Kirsch. Nonreflecting boundary condition for time-dependent multiple scattering. *J. Comput. Phys.*, 221:41–62, 2007.

37. T. Hagstrom. New results on absorbing layers and radiation boundary conditions. In *Topics in computational wave propagation*, volume 31 of *Lect. Notes Comput. Sci. Eng.*, pages 1–42. Springer, Berlin, 2003.

38. L. Halpern and J. Rauch. Error analysis for absorbing boundary conditions. *Numerische Mathematik*, 51(4):459–467, 1987.

39. W. Hunziker, I. M. Sigal, and A. Soffer. Minimal escape velocities. *Comm. Partial Differential Equations*, 24(11–12):2279–2295, 1999.

40. S. Ji, Y. Yang, G. Pang, and X. Antoine. Accurate artificial boundary conditions for the semi-discretized linear schrödinger and heat equations on rectangular domains. *Computer Physics Communications*, 222:84–93, 2018.

41. S. Jiang and L. Greengard. Efficient representation of nonreflecting boundary conditions for the time-dependent schrodinger equation in two dimensions. *Communications in Pure and Applied Mathematics*, 61(2):261–288, 2008.

42. Shidong Jiang and L. Greengard. Fast evaluation of nonreflecting boundary conditions for the Schrödinger equation in one dimension. *Comput. Math. Appl.*, 47(6–7):955–966, 2004.

43. M. Kazakova and P. Noble. Discrete transparent boundary conditions for the linearized green–naghdi system of equations. *SIAM Journal on Numerical Analysis*, 58(1):657–683, 2020.

44. J. B. Keller and D. Givoli. Exact non-reflecting boundary conditions. *Journal of computational physics*, 82(1):172–192, 1989.

45. P. Klein, X. Antoine, C. Besse, and M. Ehrhardt. Absorbing boundary conditions for solving n-dimensional stationary schrodinger equations with unbounded potentials and nonlinearities. *Communications in Computational Physics*, 10(5):1280–1304, 2011.

46. B. Li, J. Zhang, and C. Zheng. Stability and error analysis for a second-order fast approximation of the one-dimensional schrodinger equation under absorbing boundary conditions. *SIAM Journal on Scientific Computing*, 40(6):A4083–A4104, 2018.

47. X. Li. Absorbing boundary conditions for time-dependent schrödinger equations: A density-matrix formulation. *The Journal of chemical physics*, 150(11):114111, 2019.

48. B. Liu and A. Soffer. A general scattering theory for nonlinear and non-autonomous schroedinger type equations-a brief description. *arXiv preprint arXiv:2012.14382*, 2020.

49. C. Lubich and A. Schädle. Fast convolution for nonreflecting boundary conditions. *SIAM J. Sci. Comput.*, 24(1):161–182 (electronic), 2002.

50. G. Mur. Absorbing boundary conditions for the finite-difference approximation of the time-domain electromagnetic-field equations. *IEEE transactions on Electromagnetic Compatibility*, (4):377–382, 1981.

51. D. Neuhauser and M. Baer. The time-dependent schrodinger equation: Application of absorbing boundary conditions. *J. Chem. Phys.*, 90(8), 1989.

52. D. P. C. Nguyen, A. Soffer, C. Stucchio, and M.-B. Tran. A phase space filter algorithm for anisotropic waves. *In Preparation*.

53. A. Nissen and G. Kreiss. An optimized perfectly matched layer for the schrödinger equation. *Communications in Computational Physics*, 9(1):147–179, 2011.

54. S.A. Orszag. On the elimination of aliasing in finite difference schemes by filtering high-wavenumber components. *Journal of the Atmospheric Sciences*, 28:1074.

55. G. Pang, L. Bian, and S. Tang. Almost exact boundary condition for one-dimensional schrödinger equations. *Physical Review E*, 86(6):066709, 2012.

56. J. S. Papadakis, M. I. Taroudakis, Panagiotis J. Papadakis, and B. Mayfield. A new method for a realistic treatment of the sea bottom in the parabolic approximation. *The Journal of the Acoustical Society of America*, 92(4):2030–2038, 1992.

57. Z. Rapti, M. I. Weinstein, and P. G. Kevrekidis. Transient radiative behavior of Hamiltonian systems in finite domains. *Phys. Lett. A*, 345(1–3):1–9, 2005.

58. D. Gottlieb S. Abarbanel and J.S. Hesthaven. Long time behavior of the perfectly matched layer equations in computational electromagnetics. *Journal of Scientific Computing*, 17(1), 2002.

59. A. Schädle. Non-reflecting boundary condition for a Schrödinger-type equation. In *Mathematical and numerical aspects of wave propagation (Santiago de Compostela, 2000)*, pages 621–625. SIAM, Philadelphia, PA, 2000.

60. A. Schädle. Non-reflecting boundary conditions for the two-dimensional Schrödinger equation. *Wave Motion*, 35(2):181–188, 2002.

61. I. M. Sigal and A. Soffer. Asymptotic completeness of N-particle long-range scattering. *J. Amer. Math. Soc.*, 7(2):307–334, 1994.

62. A. Soffer and C. Stucchio. Open boundaries for the nonlinear schrodinger equation. *Journal of Computational Physics*, 225(2):1218–1232, 2007.

63. A. Soffer and C. Stucchio. Multiscale resolution of shortwave-longwave interactions in time dependent dispersive waves. *Communications in Pure and Applied Mathematics*, 62(1):82–124, 2009.

64. A. Soffer and M. I. Weinstein. Selection of the ground state for nonlinear Schrödinger equations. *Rev. Math. Phys.*, 16(8):977–1071, 2004.

65. C. Stucchio. *Selected Problems in Quantum Mechanics, PhD Thesis, Rutgers University*.

66. J. Szeftel. Design of absorbing boundary conditions for Schrödinger equations in $\mathbb{R}^d$. *SIAM J. Numer. Anal.*, 42(4):1527–1551 (electronic), 2004.

67. J. Szeftel. Absorbing boundary conditions for nonlinear schrodinger equations. *Numerische Mathematik*, (104):103–127, 2006.

68. L. N. Trefethen and L. Halpern. Well-posedness of one-way wave equations and absorbing boundary conditions. *Mathematics of computation*, 47(176):421–435, 1986.

69. C. Besse X. Antoine and S. Descombes. Artificial boundary conditions for one-dimensional cubic nonlinear schrodinger equations. *SIAM J. Numer. Anal.*, 43(6):2272–2293, 2006.

70. C. Besse X. Antoine and V. Mouysset. Numerical schemes for the simulation of the two-dimensional Schrödinger equation using non-reflecting boundary conditions. *Math. Comp.*, 73(248):1779–1799 (electronic), 2004.

71. Z. Xu and H. Han. Absorbing boundary conditions for nonlinear schrodinger equations. *Physical Review E*, 74(3):037704, 2006.

72. Z. Xu, H. Han, and X. Wu. Adaptive absorbing boundary conditions for schrodinger-type equations: application to nonlinear and multi-dimensional problems. *Journal of Computational Physics*, 225(2):1577–1589, 2007.

73. J. Zhang, D. Li, and X. Antoine. Efficient numerical computation of time-fractional nonlinear schrödinger equations in unbounded domain. *Communications in Computational Physics*, 25(1):218–243, 2019.

74. J. Zhang, Z. Xu, and X. Wu. Unified approach to split absorbing boundary conditions for nonlinear schrodinger equations: Two-dimensional case. *Physical Review E*, 79(4):046711, 2009.

75. C. Zheng. Exact nonreflecting boundary conditions for one-dimensional cubic nonlinear schrödinger equations. *Journal of Computational Physics*, 215(2):552–565, 2006.

76. C. Zheng. A perfectly matched layer approach to the nonlinear schrödinger wave equations. *Journal of Computational Physics*, 227(1):537–556, 2007.

Printed in the United States
by Baker & Taylor Publisher Services